Contents

(1st Edition)

Published by

BG PUBLICATION

Email: chinsuankap@gmail.com

Acknowledgments

- *I praise Almighty God for guiding me to publish this book.*
- *I would also like to thank all my family for their constant support.*
- *I appreciate Mr. Kyaw Thu for his advice.*
- *Special thanks to Mr. Maung Maung Myint Thein (RIT) for correcting and editing what was needed.*
- *I would like to thank everyone who helped me though I didn't mention their names.*

Peter Kap

LOGIC GATE

In a digital system, the demonstration of electronic circuits in blocks is called a Logic Gate. Each block type is named AND Gate, OR Gate, etc. to remember easier according to their performances. Each Gate can have one or more inputs but only one output.
Due to the common use of Binary Numerical Systems in Digital Circuits, there are two Input Logic levels such as Input Logic Level "0" (or) "Low" (or) "OFF" (or) "FALSE" and Input Logic Level "1" (or) "High" (or) "ON" (or) "TRUE".
How Logic Gates work can be described as controlling a light bulb with a switch.
For example, if the switch is "ON" state, the light will light up, and if the switch is "OFF" state, the light will not light up. Input levels are indicated by Switch ON and Switch OFF.
The light light-up and not the light-up shows the output.
Also, Logic Gates can be described by using Boolean Algebra.
There are seven basic logic gates. They are AND GATE, OR GATE, NOT GATE, NAND GATE, NOR GATE, EXOR GATE, and EXNOR.

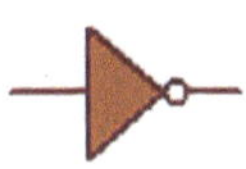

Symbol of NOT GATE

TRUTH TABLE

The table used for Boolean Algebra that describes how a Logic Gate works is called a Truth Table. The Truth Table consists of two vertical columns. They are the inputs column and the output column. The inputs column is additionally divided into sub-column according to the number of input levels that can be paired. The output results are seen depending on the combination of input levels.

INPUT		OUTPUT
A	B	$\overline{A + B}$
0	0	1
0	1	0
1	0	0
1	1	0

Truth Table of NOR GATE

BOOLEAN ALGEBRA

Boolean Algebra is a branch of mathematics used to demonstrate digital circuits mathematically.

$$Y = \overline{A \cdot B}$$

Boolean Expression of NAND GATE

PROJECT (1)

AND GATE

AND GATE

A gate that multiplies all the input levels to make a single output result is called an AND gate. An AND Gate can have two or more input levels. In the AND Gate, if any of the Input levels is "Low" or "0", the Output also becomes "Low" or "0". If all Input levels are "High" or "1", the Output result is "High" or "1".

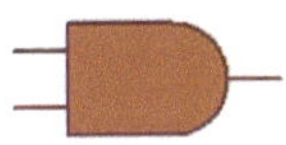

Symbol of 2-input AND GATE

$$Y = A . B$$

Boolean Expression of AND GATE

INPUT		OUTPUT
A	B	A . B
0	0	0
0	1	0
1	0	0
1	1	1

Truth Table of AND GATE

Demonstration of Truth Table

1. If input A is Logic "0" and input B is Logic "0", the light will not light up as the circuit is an open circuit.

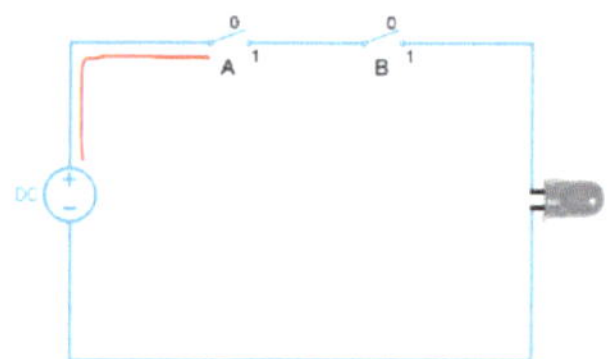

2. If input A is Logic "0" and input B is Logic "1", the light will not light up as the circuit is an open circuit.

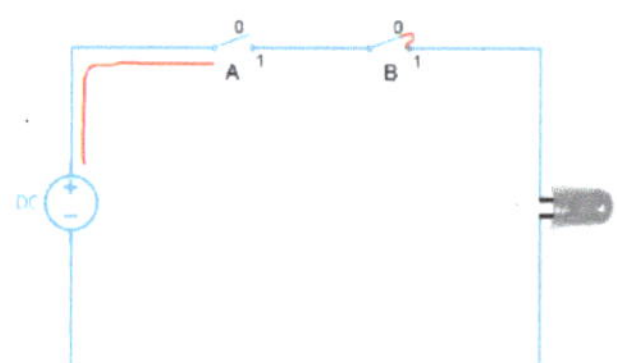

3. If input A is Logic "1" and input B is Logic "0", the light will not light up as the circuit is an open circuit.

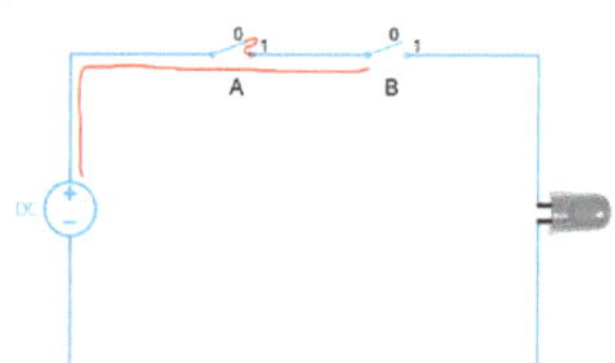

4. If input A is Logic "1", and input B is Logic "1", the light will light up as the circuit is complete.

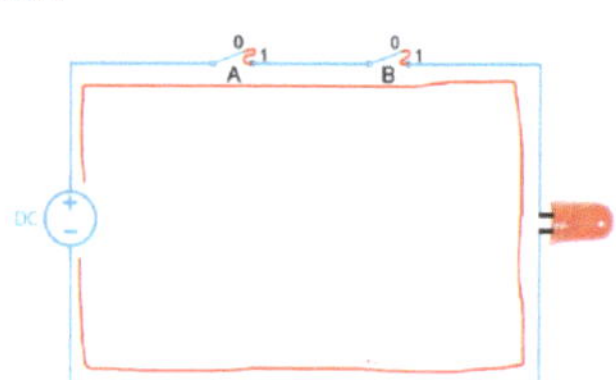

COMPONENTS

01
Push-On Switch
2 nos.

02
Transistor
2 nos.

03
Resistor 10K ohm
2 nos.

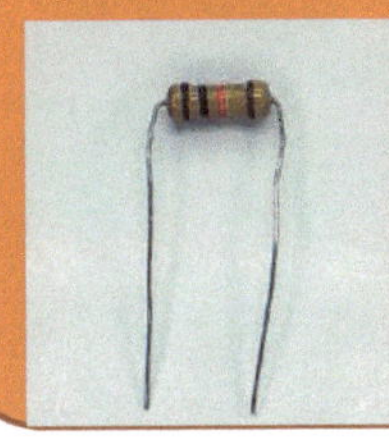

04
Resistor 1K ohm
1 no.

05
Resistor 300 ohm
1 no.

06
LED
1 no.

07
Wire
9 nos.

08
Battery 9V
1 no.

09
Connector
1 no.

AND GATE

Steps of Implementation

1.1: Place the collector pin of a transistor at the meeting point of i and 9 and the base pin of a transistor at the meeting point of i and 10, and the emitter pin of a transistor into the intersection of i and 11.

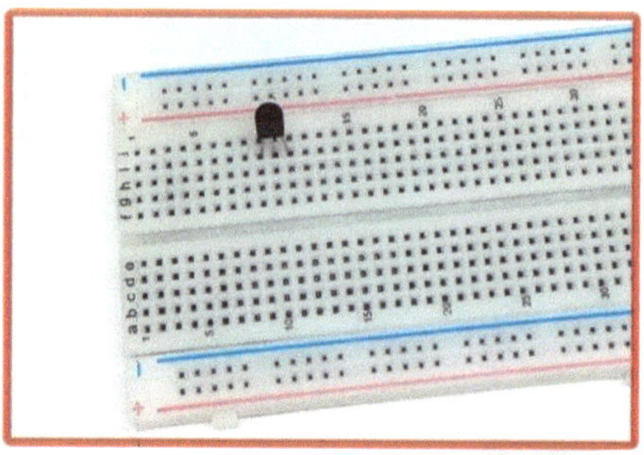

1.1

1.2: Place the collector pin of a transistor at the meeting point of j and 29 and the base pin of a transistor at the meeting point of j and 30, and the emitter pin of a transistor into the intersection of j and 31.

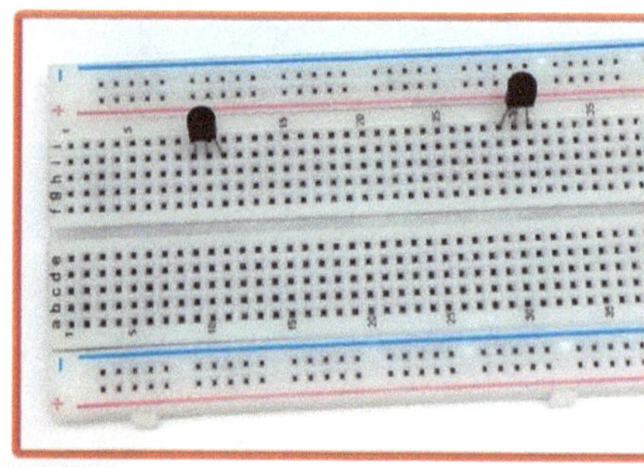

1.2

1.3

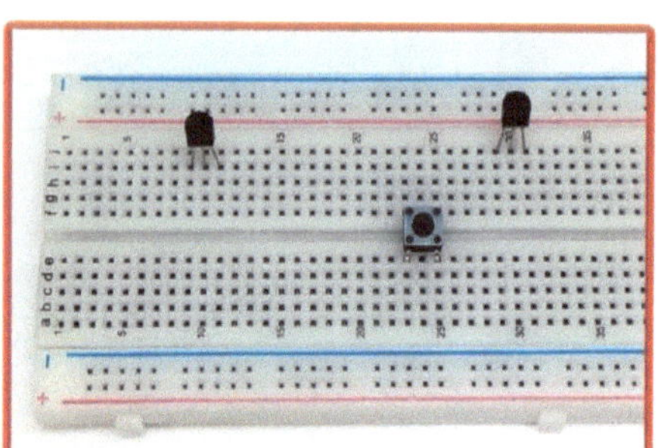

1.3: Place one of the upper two pins of a Push-On Switch where f and 23 meet and the pin next to it where f and 25 intersect. Then place one of the lower two pins of that push-on switch at the meeting point of e and 23 and the pin next to it at the meeting point of e and 25.

1.4

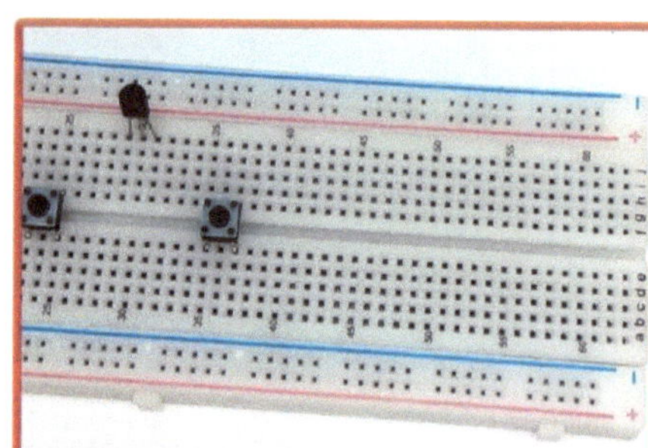

1.4: Place one of the upper two pins of a Push-On Switch where f and 35 meet and the pin next to it where f and 37 intersect. Then place one of the lower two pins of that push-on switch at the meeting point of e and 35 and the pin next to it at the meeting point of e and 37.

1.5: Put one end of a 10k Ohm Resistor pin at the meeting point of g and 10 and the rest pin of it at the meeting point of g and 25.

1.5

1.6: Put one end of a 10k Ohm Resistor pin at the meeting point of i and 30 and the rest pin of it at the meeting point of i and 35.

1.6

1.7

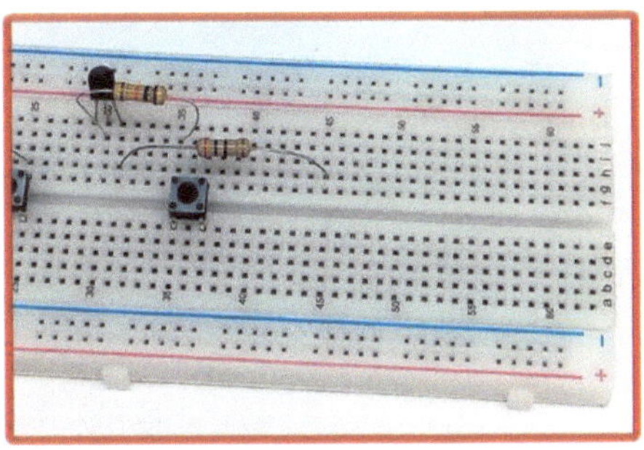

1.7: Put one end of a 300 Ohm Resistor pin at the meeting point of g and 31 and the rest pin of it at the meeting point of g and 45.

1.8

1.8: Put one end of a 1k Ohm Resistor pin into the junction of f and 31 and the rest pin of it at the negative rail on the breadboard.

AND GATE

1.9: Put the positive pin of
the blue LED
at the meeting point of
f and 45
and the negative pin of the
blue LED
into the convergence of
e and 45.

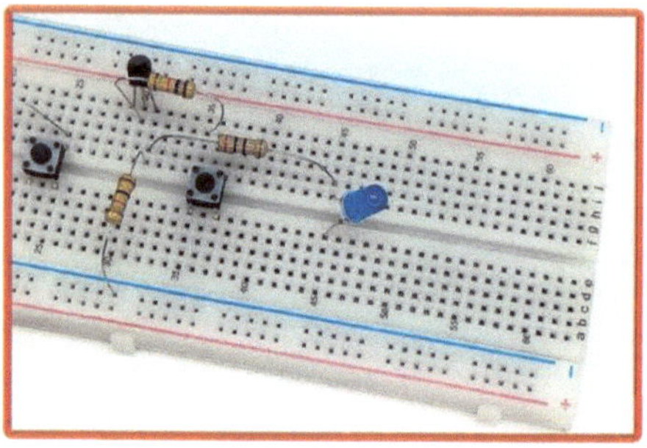

1.9

1.10: Insert one end of
a wire
into the meeting point of
j and 9
and the other end into
the positive rail
on the breadboard.

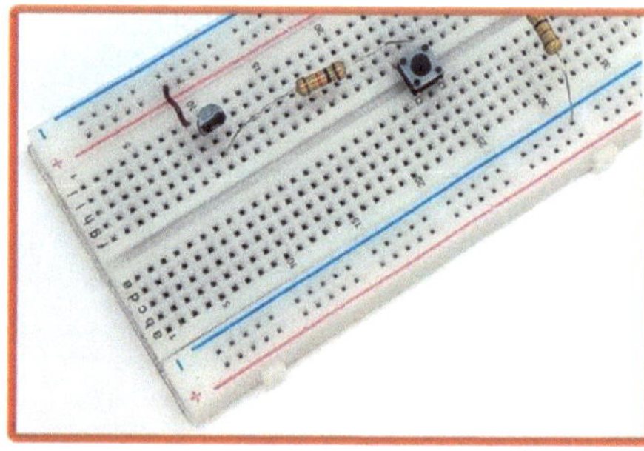

1.10

1.11

1.11: Insert one end of
a wire
into the meeting point of
j and 23
and the other end into
the positive rail
on the breadboard.

1.12

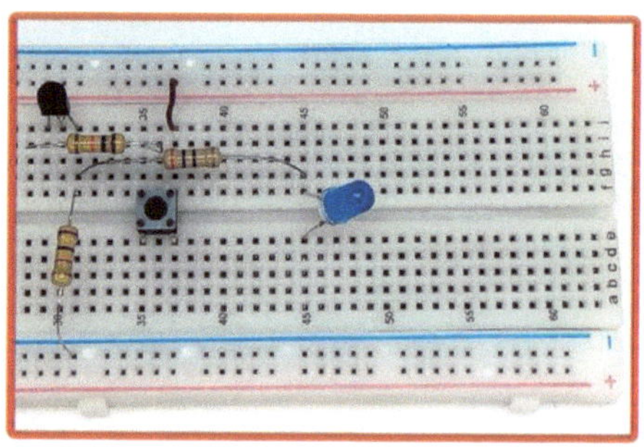

1.12: Insert one end of
a wire
into the meeting point of
j and 37
and the other end into
the positive rail
on the breadboard.

1.13: Insert one end of
a wire
into the meeting point of
a and 45
and the other end into
the negative rail
on the breadboard.

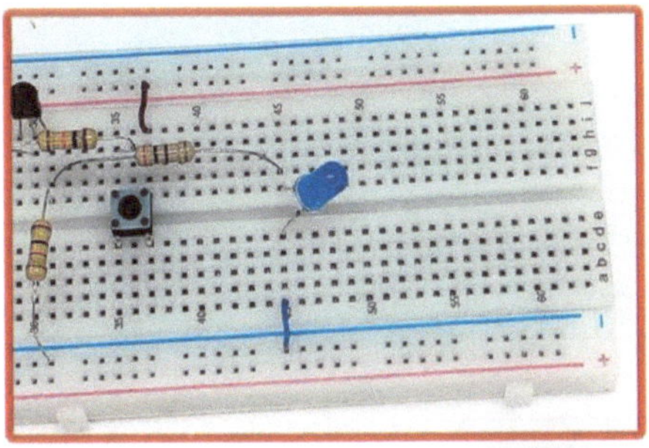

1.13

1.14: Insert one end of
a wire
into the meeting point of
h and 11
and the other end into
the meeting point of
h and 29.

1.14

1.15

1.15: Connect a 9V battery to the battery connector, and then connect the positive wire of the battery connector to the positive rail on the breadboard and the negative wire of the battery connector to the negative rail on the breadboard.

Explanation of AND GATE

Figures (1.1) to (1.9) are the placement of components.
Figures (1.10) to (1.15) are the figures of wiring the components.
The two transistors are connected in series.
The LED is connected to the emitter of the second transistor and grounded through it.
According to the truth table of the AND gate, the LED will light up only when both switches are simultaneously "ON" state.

PROJECT (2)

OR GATE

OR GATE

A gate that combines all the inputs to produce a single output is called an OR gate. An OR Gate can have two or more input levels. In the OR gate, if one of the input levels is "High" or "1", the output also becomes "High" or "1". If all Input levels are "Low" or "0", the output will be "Low" or "0".

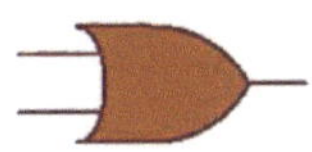

Symbol of 2-input OR GATE

$$Y = A + B$$

Boolean Expression of OR GATE

INPUT		OUTPUT
A	B	A + B
0	0	0
0	1	1
1	0	1
1	1	1

Truth Table of OR GATE

OR GATE

Demonstration of Truth Table

1. If input A is Logic "0" and input B is Logic "0", the light will not light up as the circuit is an open circuit.

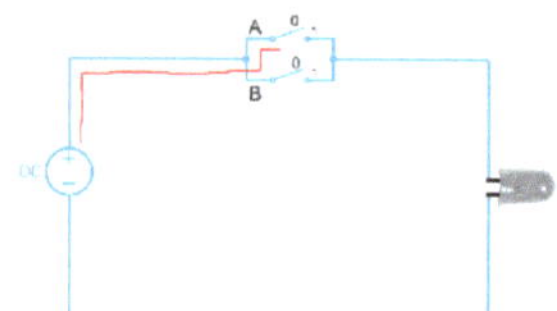

2. If input A is Logic "0", and input B is Logic "1", the light will light up as the circuit is complete.

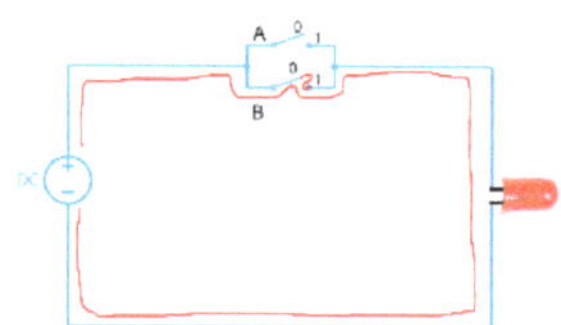

3. If input A is Logic "1", and input B is Logic "0", the light will light up as the circuit is complete.

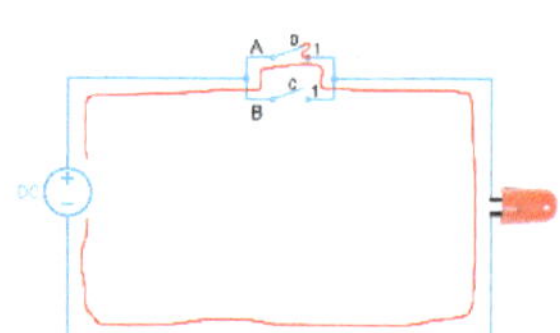

4. If input A is Logic "1", and input B is Logic "1", the light will light up as the circuit is complete.

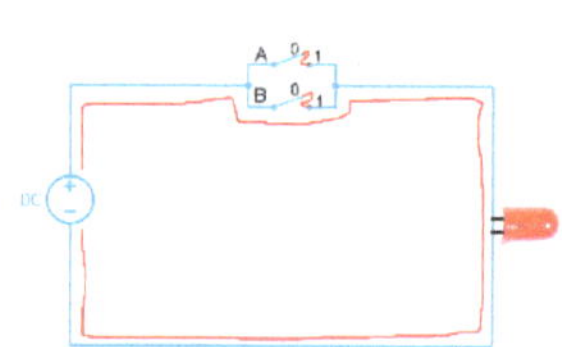

COMPONENTS

01

Push-On Switch
2 nos.

02

Transistor
2 nos.

03

Resistor 10K ohm
2 nos.

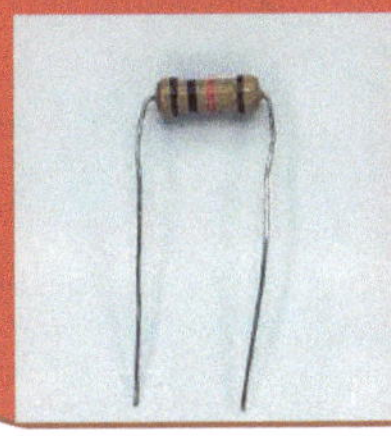

04

Resistor 1K ohm
1 no.

05

LED
1 no.

06

Wire
7 nos.

07

Battery 9V
1 no.

08

Connector
1 no.

OR GATE

Steps of Implementation

2.1: Place one of the upper two pins of a Push-On Switch where f and 10 meet and the pin next to it where f and 12 intersect. Then place one of the lower two pins of that push-on switch at the meeting point of e and 10 and the pin next to it at the meeting point of e and 12.

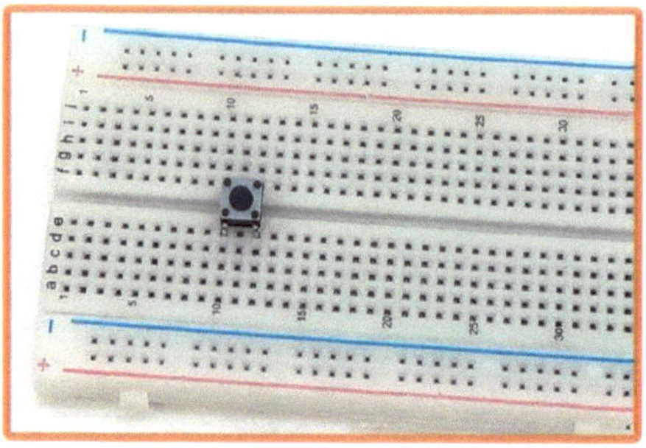

2.1

2.2: Place one of the upper two pins of a Push-On Switch where f and 18 meet and the pin next to it where f and 20 intersect. Then place one of the lower two pins of that push-on switch at the meeting point of e and 18 and the pin next to it at the meeting point of e and 20.

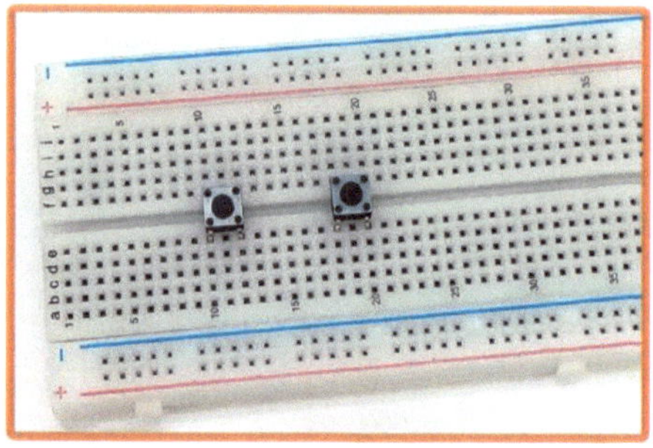

2.2

2.3

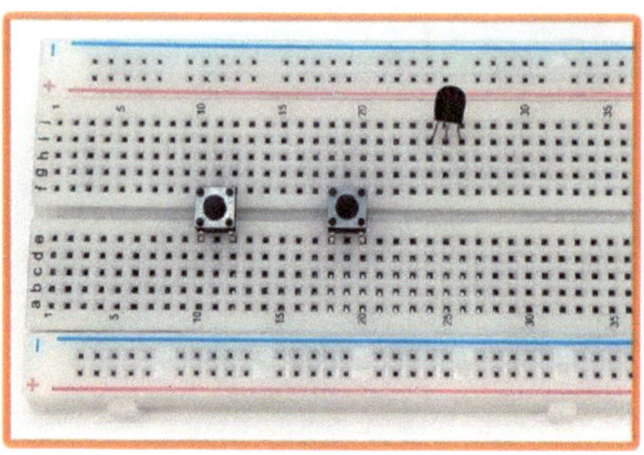

2.3: Place the collector pin of a transistor at the meeting point of i and 24 and the base pin of a transistor at the meeting point of i and 25, and the emitter pin of a transistor into the intersection of i and 26.

2.4

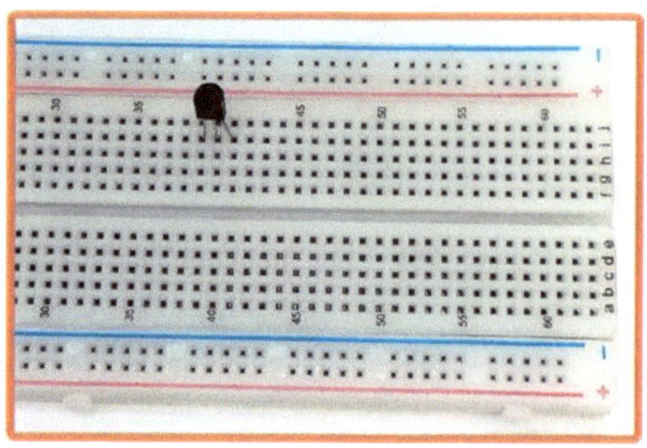

2.4: Place the collector pin of a transistor at the meeting point of i and 39 and the base pin of a transistor at the meeting point of i and 40, and the emitter pin of a transistor into the intersection of i and 41.

2.5: Put one end of a 10k Ohm Resistor pin at the meeting point of h and 12 and the rest pin of it at the meeting point of h and 25.

2.6: Put one end of a 10k Ohm Resistor pin at the meeting point of h and 35 and the rest pin of it at the meeting point of h and 40.

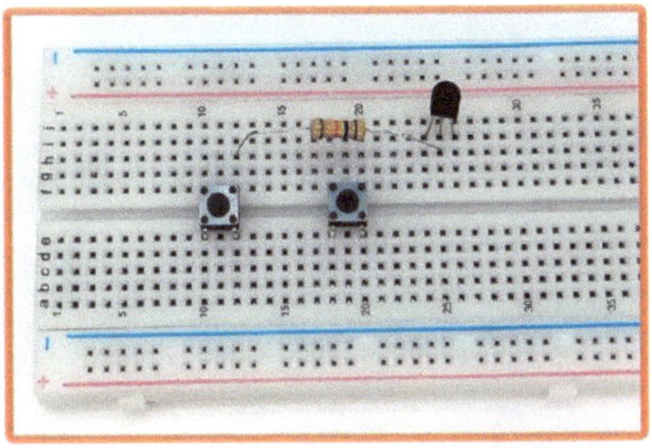

2.5

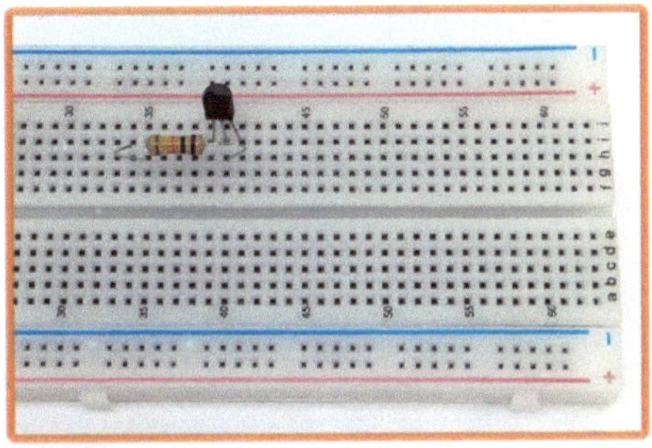

2.6

2.7

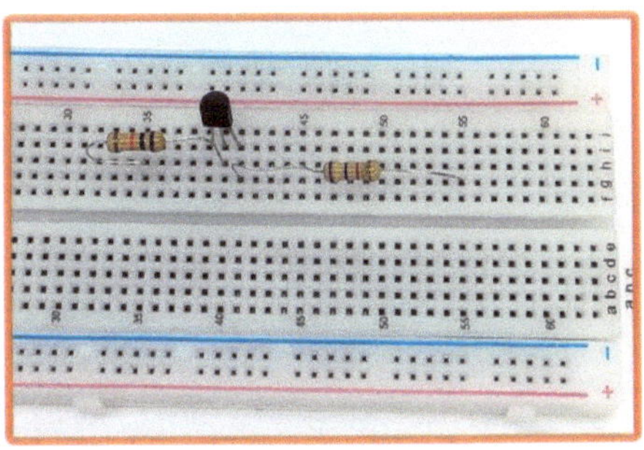

2.8

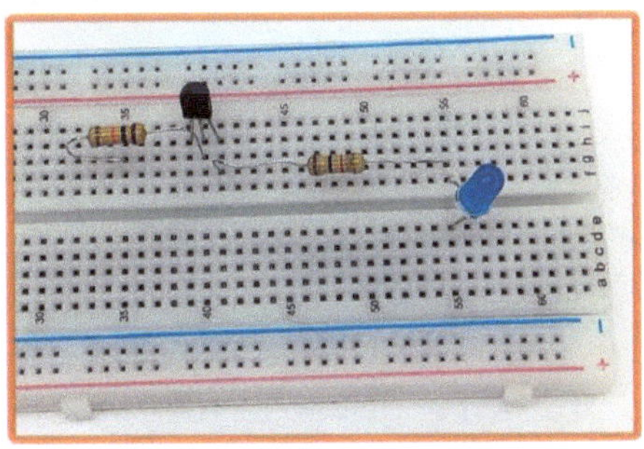

2.7: Put one end of a 1k Ohm Resistor pin at the meeting point of g and 41 and the rest pin of it at the meeting point of g and 55.

2.8: Put the positive pin of the blue LED at the meeting point of f and 55 and the negative pin of the blue LED at the junction of e and 55 on the breadboard.

OR GATE

2.9: Insert one end of a wire into the meeting point of j and 10 and the other end into the positive rail on the breadboard.

2.9

2.10: Insert one end of a wire into the meeting point of j and 18 and the other end into the positive rail on the breadboard.

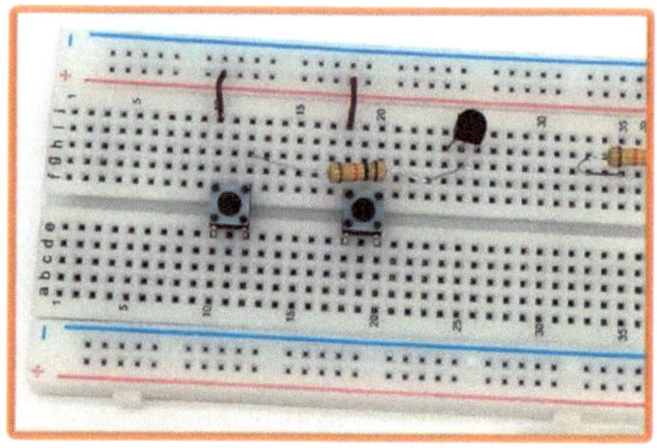

2.10

2.11

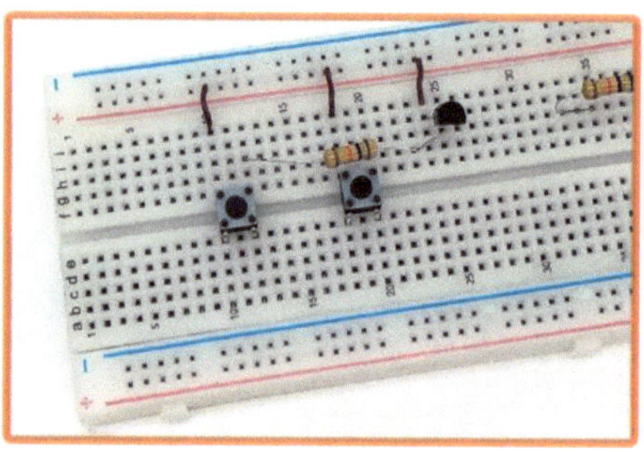

2.11: Insert one end of a wire into the meeting point of j and 24 and the other end into the positive rail on the breadboard.

2.12

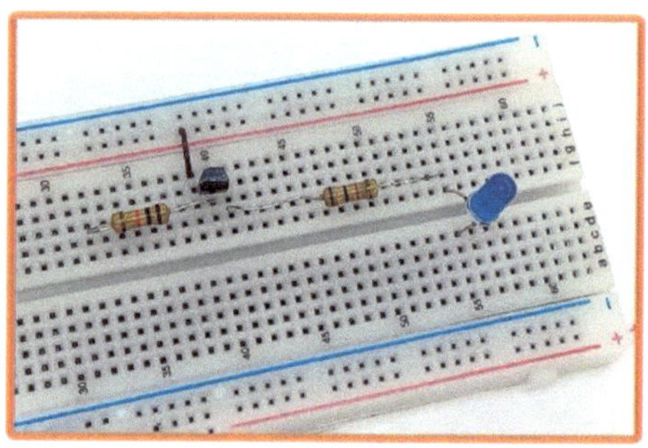

2.12: Insert one end of a wire into the meeting point of j and 39 and the other end into the positive rail on the breadboard.

2.13: Insert one end of a wire into the meeting point of g and 20 and the other end into the meeting point of g and 35.

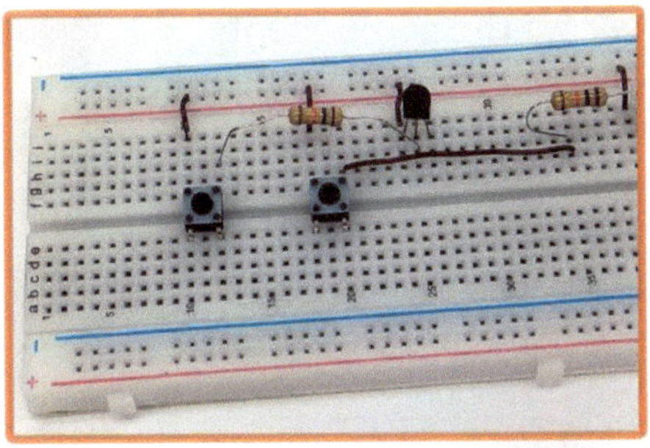

2.13

2.14: Insert one end of a wire into the meeting point of f and 26 and the other end into the meeting point of f and 41.

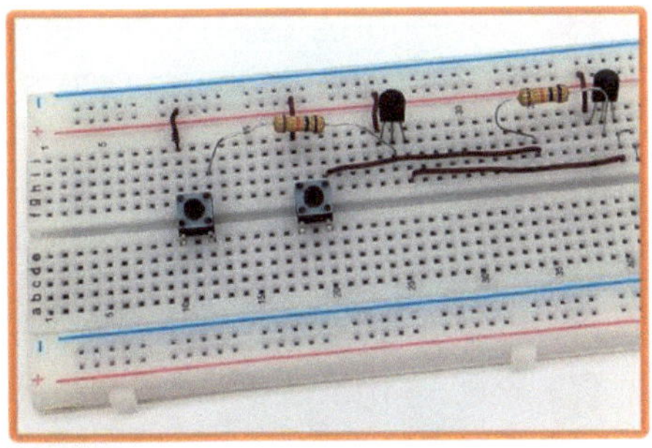

2.14

2.15

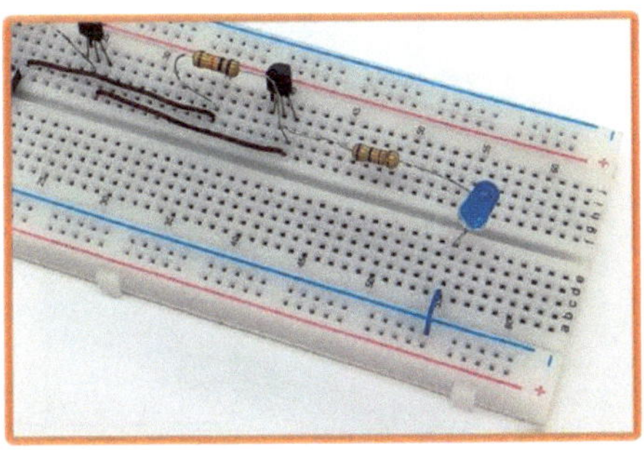

2.15: Insert one end of a wire into the meeting point of a and 55 and the other end into the negative rail on the breadboard.

2.16

2.16: Connect a 9V battery to the battery connector, and then connect the positive wire of the battery connector to the positive rail on the breadboard and the negative wire of the battery connector to the negative rail on the breadboard.

Explanation of OR GATE

Figures (2.1) to (2.8) are the placement of components.
Figures (2.9) to (2.16) are wiring the components.
The two transistors are connected in parallel.
The LED is connected to the emitter of the second transistor
and grounded through it.
According to the truth table of the OR gate,
if both switches are "OFF" state at the same time,
the LED will not light up. In other conditions,
the LED will light up.

PROJECT (3)

NOT GATE

NOT GATE

A gate whose output is the opposite or inverse of the input is called a NOT gate. A NOT gate has only one input level. NOT Gate is also called Inverter Gate. If the input level is "Low" or "0", the output is "High" or "1". The small circle at the output of the NOT Gate symbol is called a "Bubble." It symbolizes the Logic Level to be inverted or the opposite.

INPUT	OUTPUT
A	$\bar{A}$
0	1
1	0

Truth Table of NOT GATE

Symbol of NOT GATE

$$Y = \bar{A}$$

Boolean Expression of NOT GATE

Demonstration of Truth Table

1. If input A is Logic "0" the light will light up as the circuit is complete.

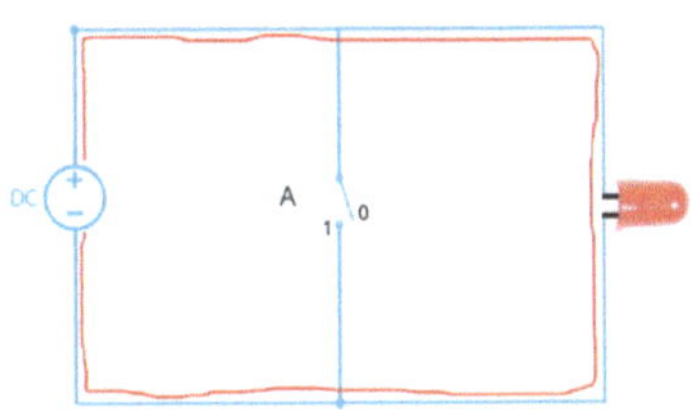

2. If input A is Logic "1" the light will not light up as the circuit is a Short Circuit.

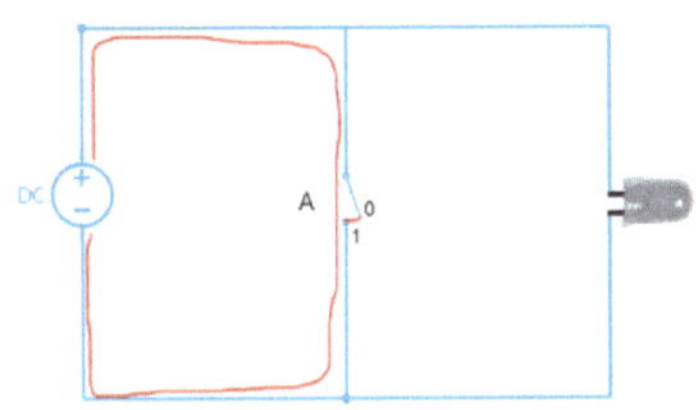

COMPONENTS

01

Push-On Switch
1 no.

02

Transistor
1 no.

03

Resistor 10K ohm
2 nos.

04

LED
1 no.

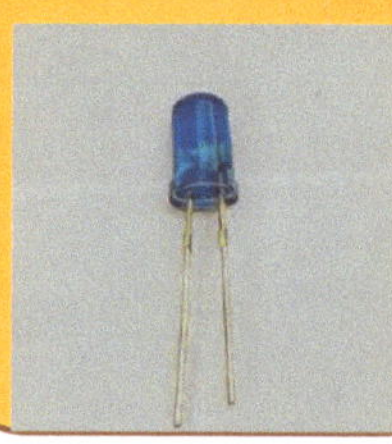

05

Wire
5 nos.

06

Battery 9V
1 no.

07

Connector
1 no.

Steps of Implementation

3.1: Place one of the upper two pins of a Push-On Switch where f and 15 meet and the pin next to it where f and 17 intersect. Then place one of the lower two pins of that push-on switch at the meeting point of e and 15 and the pin next to it at the meeting point of e and 17.

3.1

3.2: Place the collector pin of a transistor at the meeting point of i and 34 and the base pin of a transistor at the meeting point of i and 35, and the emitter pin of a transistor into the intersection of i and 36.

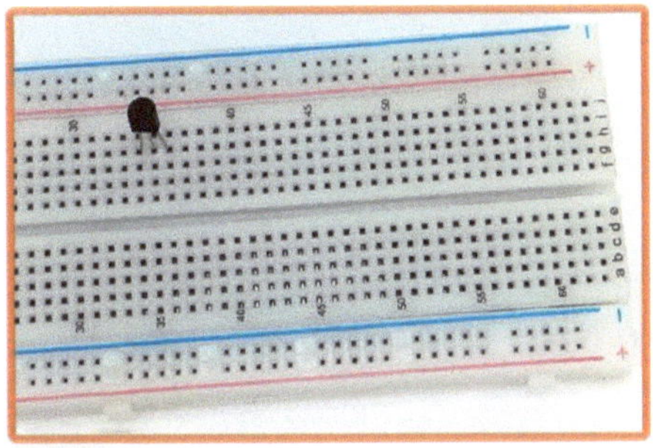

3.2

3.3

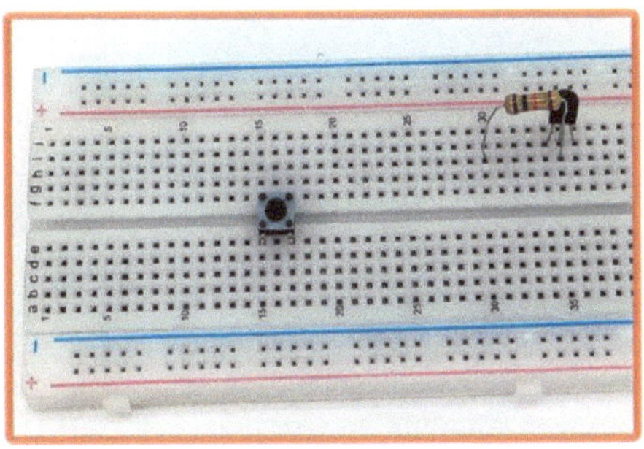

3.3: Put one end of a 10k Ohm Resistor pin at the meeting point of h and 30 and the rest pin of it at the meeting point of h and 35.

3.4

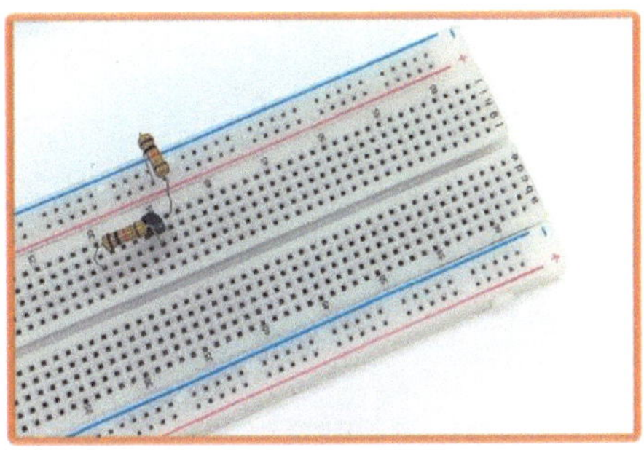

3.4: Put one end of a 10k Ohm Resistor pin into the junction of j and 36 and the rest pin of it at the positive rail on the breadboard.

3.5: Put the positive pin of the blue LED bulb at the meeting point of f and 40 and the negative pin of the blue LED bulb into the convergence of e and 40.

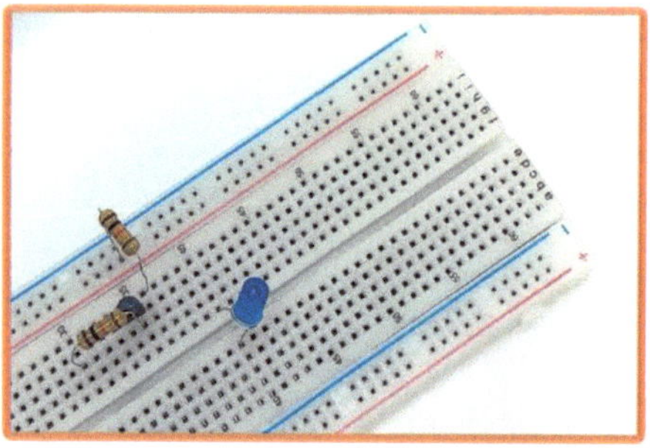

3.5

3.6: Insert one end of a wire into the meeting point of j and 15 and the other end into the positive rail on the breadboard.

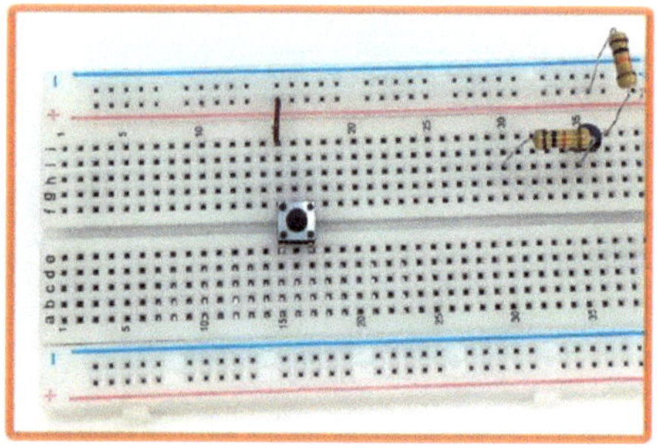

3.6

3.7

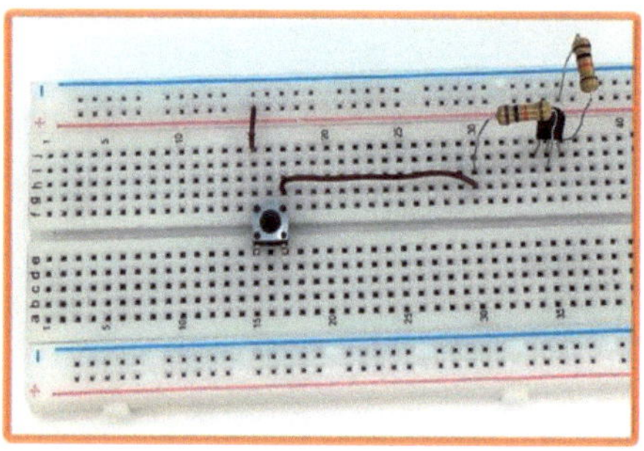

3.7: Insert one end of a wire into the meeting point of g and 17 and the other end into the meeting point of g and 30.

3.8

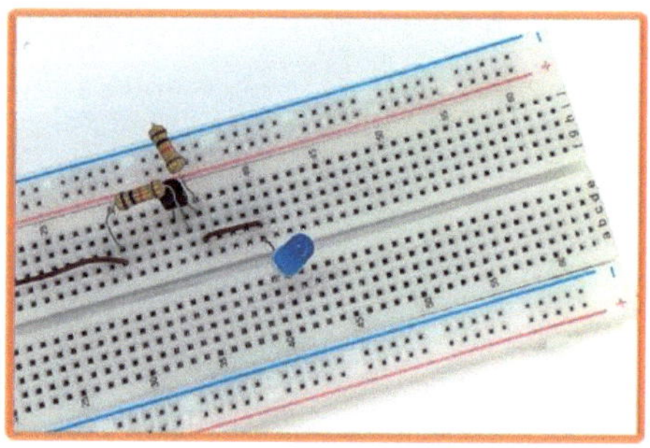

3.8: Insert one end of a wire into the meeting point of g and 36 and the other end into the meeting point of g and 40.

3.9: Insert one end of a wire into the meeting point of g and 20 and the other end into the meeting point of g and 35.

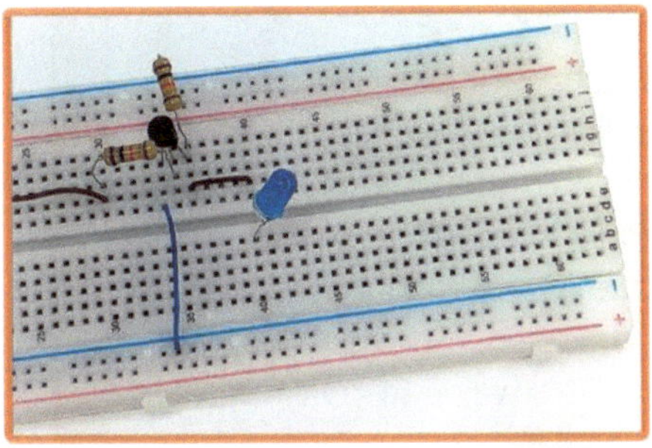

3.9

3.10: Insert one end of a wire into the meeting point of f and 26 and the other end into the meeting point of f and 41.

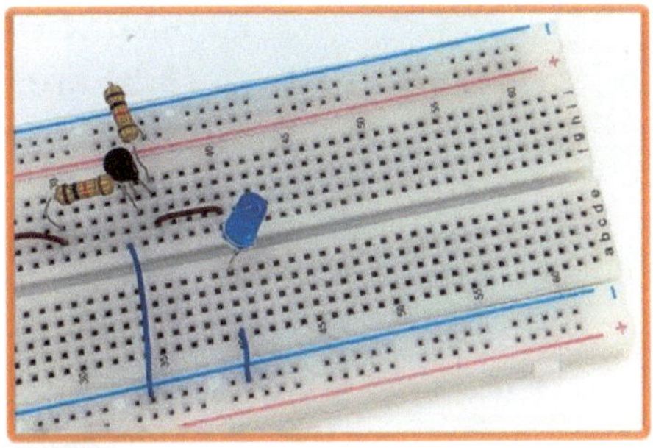

3.10

3.11

3.11: Connect a 9V battery to the battery connector, and then connect the positive wire of the battery connector to the positive rail on the breadboard and the negative wire of the battery connector to the negative rail on the breadboard.

Explanation of NOT GATE

Figures (3.1) to (3.5) are the placement of components. Figures (3.6) to (3.11) are wiring the figures of the components. The LED is connected to the collector of the transistor and grounded through it.
According to the truth table of the NOT gate, when the switch is in an "OFF" state, the LED bulb will light up. When the switch is "ON" state, the LED bulb will not light up.

NAND GATE

NAND GATE

A gate whose output is the opposite or inverse result of multiplying all input levels is called a NAND gate. A NAND Gate can have two or more inputs. A NAND Gate is obtained by combining an AND Gate and a NOT Gate. In NAND Gate, if one of the input levels is "Low" or "0", the output becomes "High" or "1". If all Input levels are "High" or "1", the output will be "Low" or "0".

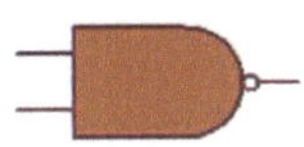

Symbol of 2-input NAND GATE

$$Y = \overline{A . B}$$

w

Boolean Expression of NAND GATE

INPUT		OUTPUT
A	B	$\overline{A . B}$
0	0	1
0	1	1
1	0	1
1	1	0

Truth Table of NAND GATE

NAND GATE

Demonstration of Truth Table

1. If input A is Logic "0", and input B is Logic "0", the light will light up as the circuit is complete.

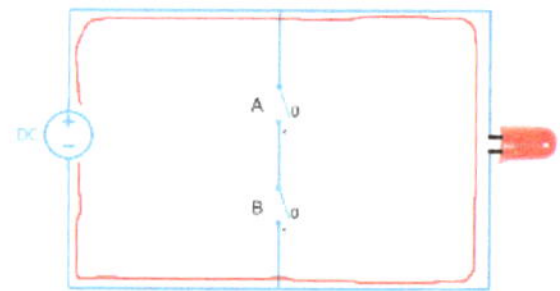

2. If input A is Logic "0", and input B is Logic "1", the light will light up as the circuit is complete.

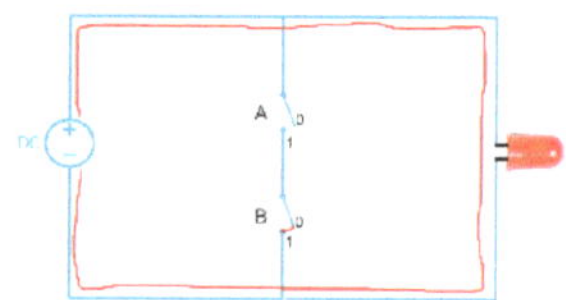

3. If input A is Logic "1", and input B is Logic "0", the light will light up as the circuit is complete.

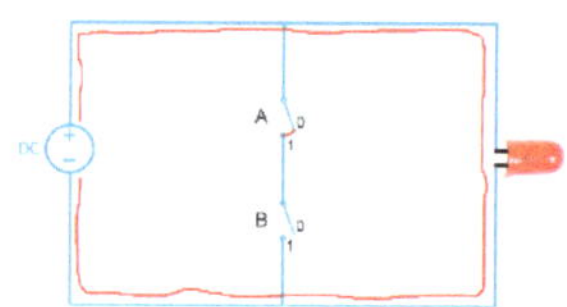

4. If input A is Logic "1" and input B is Logic "1", the light will not light up as the circuit is a Short Circuit.

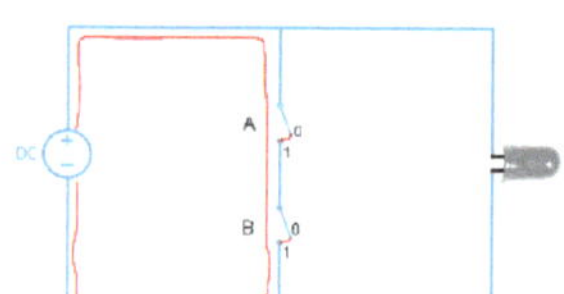

COMPONENTS

01

Push-On Switch
2 nos.

02

Transistor
2 nos.

03

Resistor 10K ohm
2 nos.

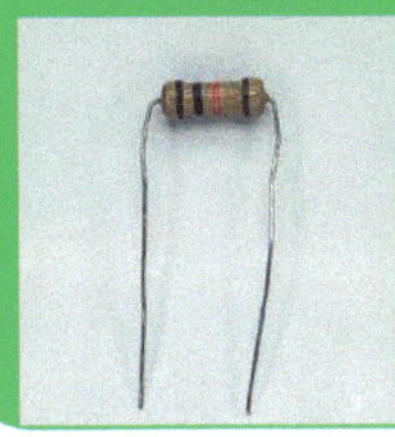

04

Resistor 1K Ohm
1 no.

05

LED
1 no.

06

Wire
6 nos.

07

Battery 9V
1 no.

08

Connector
1 no.

NAND GATE

Steps of Implementation

4.1: Place one of the upper two pins of a Push-On Switch where f and 13 meets and the pin next to it where f and 15 intersect. Then place one of the lower two pins of that push-on switch at the meeting point of e and 13 and the pin next to it at the meeting point of e and 15.

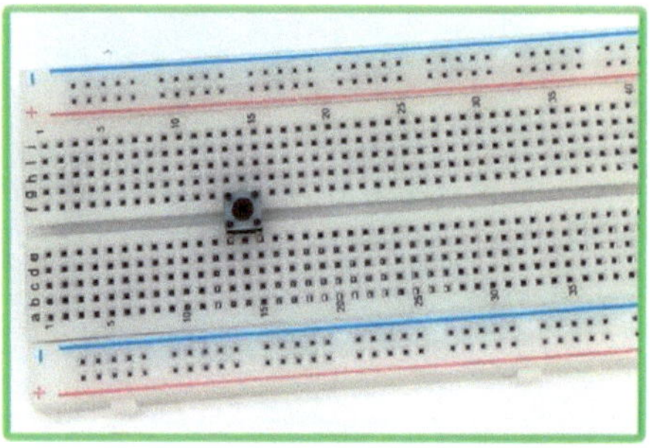

4.1

4.2: Place one of the upper two pins of a Push-On Switch where f and 25 meets and the pin next to it where f and 27 intersect. Then place one of the lower two pins of that push-on switch at the meeting point of e and 25 and the pin next to it at the meeting point of e and 27.

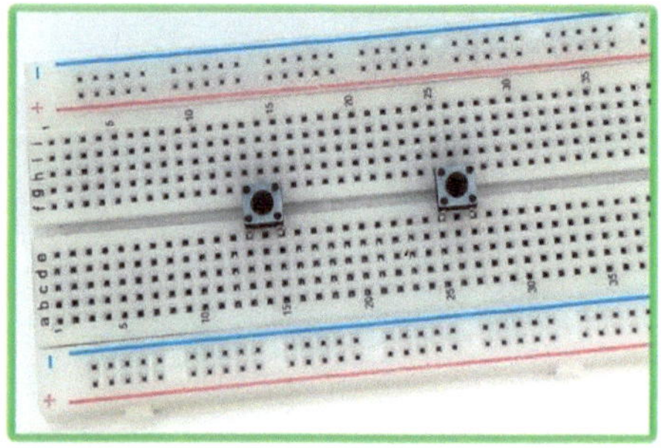

4.2

4.3

4.3: Put one end of a 10k Ohm Resistor pin at the meeting point of h and 15 and the rest pin of it at the meeting point of h and 20.

4.4

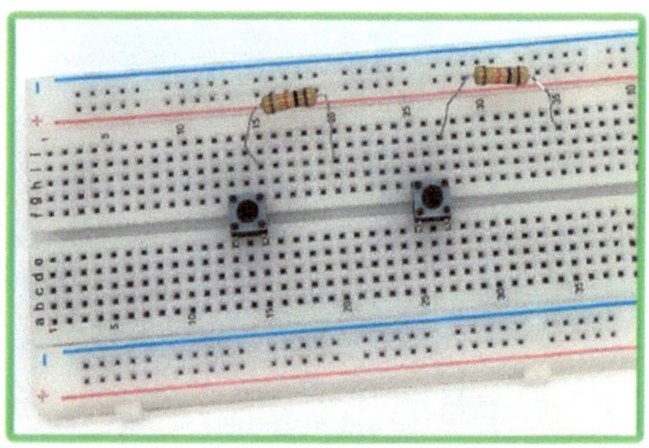

4.4: Put one end of a 10k Ohm Resistor pin at the meeting point of i and 27 and the rest pin of it at the meeting point of i and 35.

4.5: Put one end of a 1k Ohm Resistor pin into the junction of j and 55 and the rest pin of it at the positive rail on the breadboard.

4.6: Place the collector pin of a transistor at the meeting point of j and 34 and the base pin of a transistor at the meeting point of j and 35, and the emitter pin of a transistor into the intersection of j and 36.

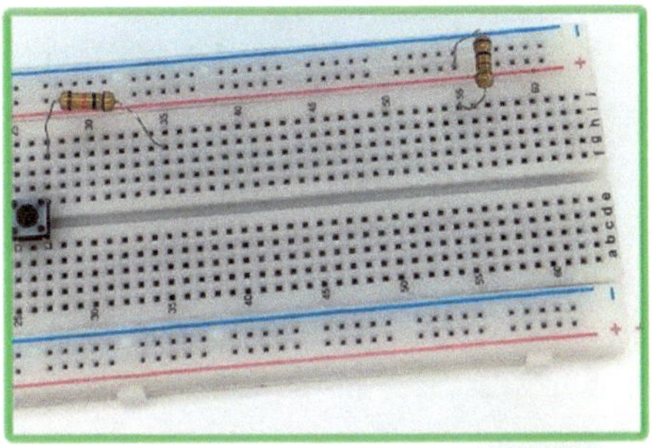

4.5

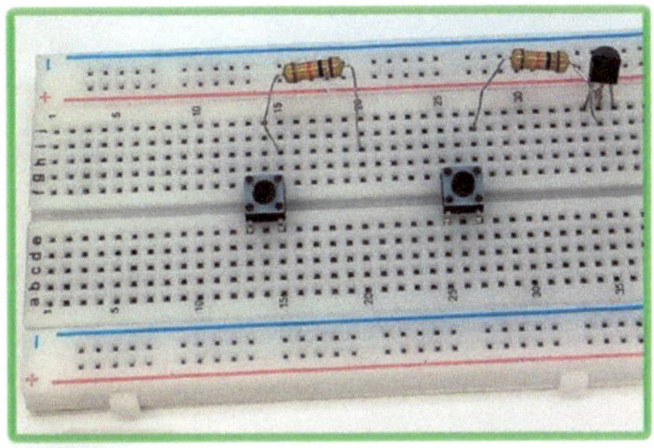

4.6

4.7

4.8

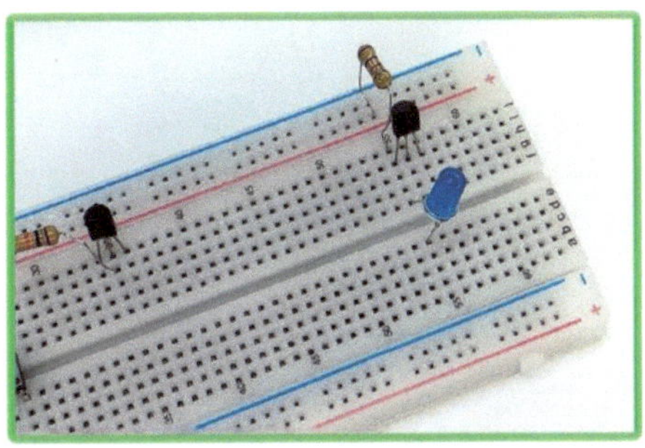

4.7: Place the collector pin of a transistor at the meeting point of i and 55 and the base pin of a transistor at the meeting point of i and 56, and the emitter pin of a transistor into the intersection of i and 57.

4.8: Put the positive pin of the blue LED at the meeting point of f and 55 and the negative pin of the blue LED into the convergence of e and 55.

4.9: Insert one end of
a wire into
the meeting point of
j and 13
and the other end into
the positive rail
on the breadboard.

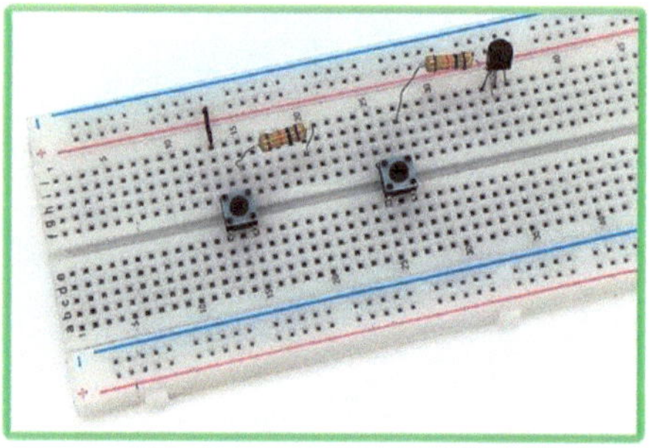

4.9

4.10: Insert one end of
a wire into
the meeting point of
j and 25
and the other end into
the positive rail
on the breadboard.

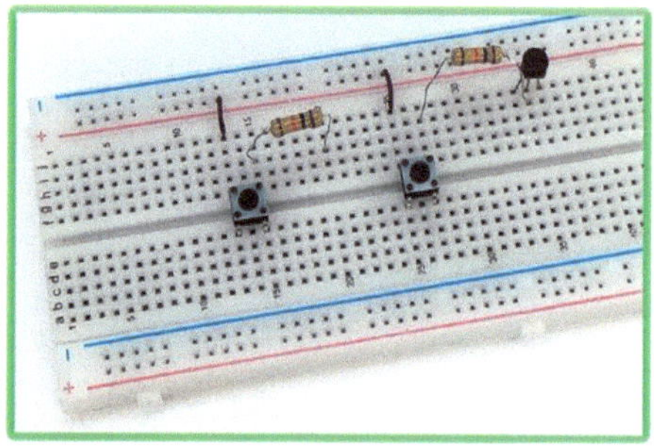

4.10

4.11

4.11: Insert one end of
a wire into
the meeting point of
g and 20
and the other end into
the meeting point of
g and 56.

4.12

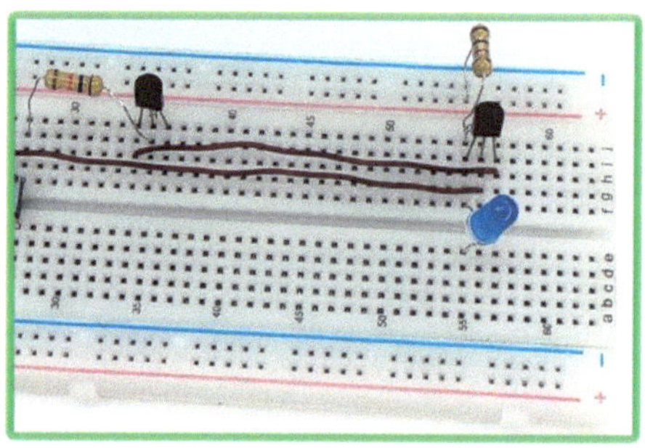

4.12: Insert one end of
a wire into
the meeting point of
h and 34
and the other end into
the meeting point of
h and 57.

4.13: Insert one end of
a wire into
the meeting point of
f and 36
and the other end into
the negative rail
on the breadboard.

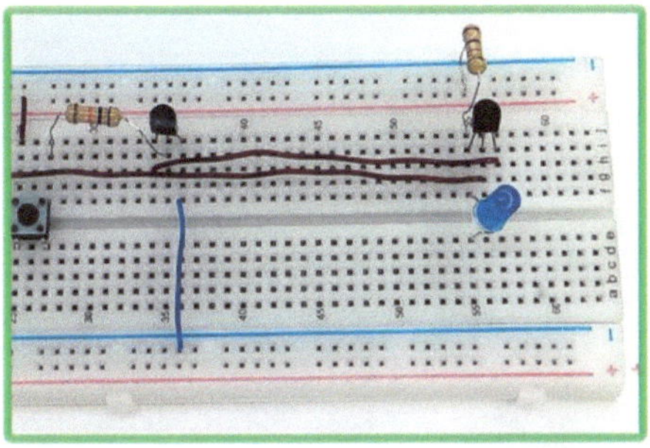

4.13

4.14: Insert one end of
a wire into
the meeting point of
a and 55
and the other end into
the negative rail
on the breadboard.

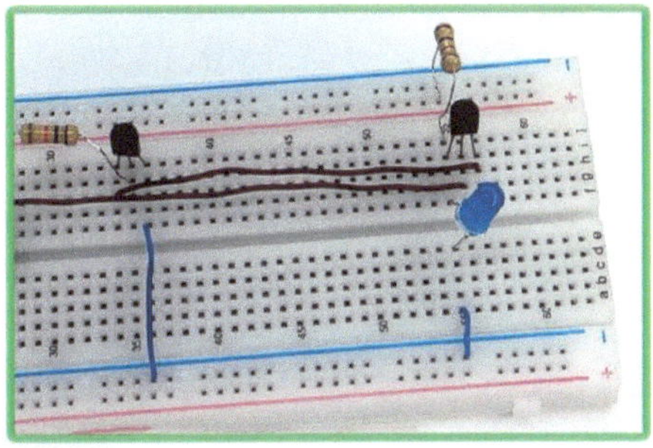

4.14

4.15

4.15: Connect a 9V battery to the
battery connector,
and then connect the positive wire of
the battery connector to
the positive rail on the breadboard
and the negative wire of the battery
connector to
the negative rail on the breadboard.

Explanation of NAND GATE

Figures (4.1) to (4.8) are the placement of components.
Figures (4.9) to (4.15) are the figures of wiring components.
The two transistors are connected in series.
The LED is connected to the collector of the first transistor and grounded through it.
According to the truth table of the NAND gate, if both switches are "ON" state at the same time, LEDs will turn off. In other conditions, the LED light will be on.

PROJECT (5)

NOR GATE

NOR GATE

A gate whose output is the opposite or inverse of the sum of all input levels is called a NOR gate. A NOR Gate can have two or more inputs. A NOR gate is obtained by combining an OR gate and a NOT gate. In a NOR gate, if one of the input levels is "High" or "1", the output becomes "Low" or "0". If all Input levels are "Low" or "0", the output will be "High" or "1".

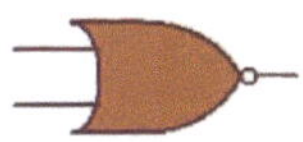

Symbol of 2-input NOR GATE

$$Y = \overline{A + B}$$

Boolean Expression of NOR GATE

INPUT		OUTPUT
A	B	$\overline{A + B}$
0	0	1
0	1	0
1	0	0
1	1	0

Truth Table of NOR GATE

NOR GATE

Demonstration of Truth Table

1. If input A is Logic "0", and input B is Logic "0", the light will light up as the circuit is complete.

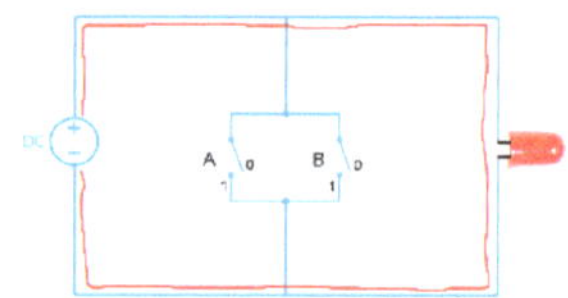

2. If input A is Logic "0" and input B is Logic "1", the light will not light up as the circuit is a Short Circuit.

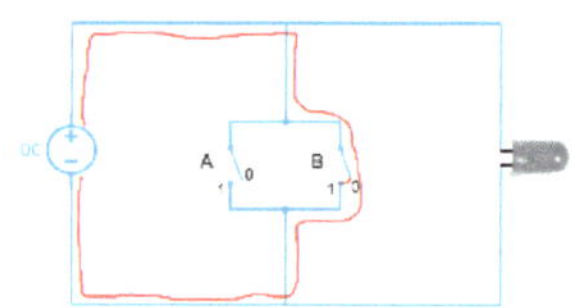

3. If input A is Logic "1" and input B is Logic "0", the light will not light up as the circuit is a Short Circuit.

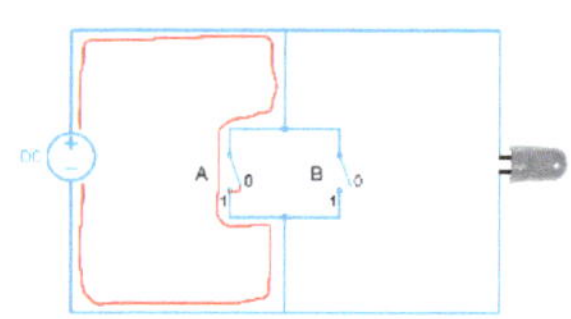

4. If input A is Logic "1" and input B is Logic "1", the light will not light up as the circuit is a Short Circuit.

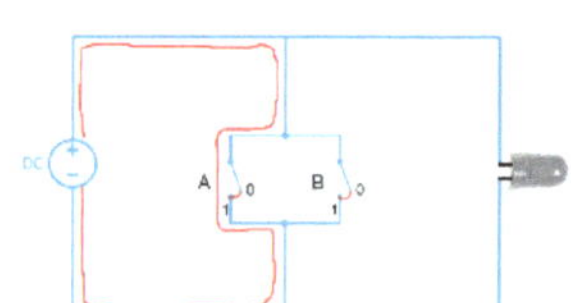

COMPONENTS

01

Push-On Switch
2 nos.

02

Transistor
2 nos.

03

Resistor 10K ohm
2 nos.

04

Resistor 1K Ohm
2 nos.

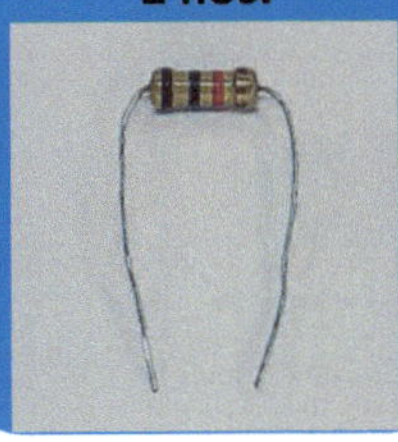

05

LED
1 no.

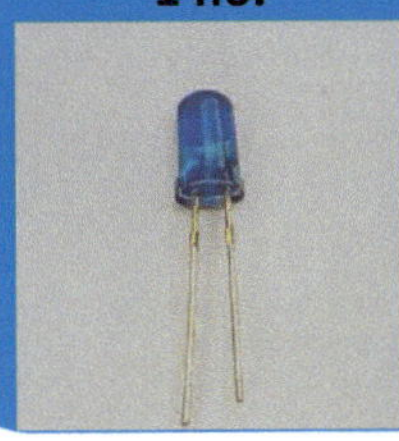

06

Wire
8 nos.

07

Battery 9V
1 no.

08

Connector
1 no.

NOR GATE

Steps of Implementation

5.1: Place one of the upper two pins of a Push-On Switch where f and 10 meets and the pin next to it where f and 12 intersect. Then place one of the lower two pins of that push-on switch at the meeting point of e and 10 and the pin next to it at the meeting point of e and 12.

5.2: Place one of the upper two pins of a Push-On Switch where f and 45 meets and the pin next to it where f and 47 intersect. Then place one of the lower two pins of that push-on switch at the meeting point of e and 45 and the pin next to it at the meeting point of e and 47.

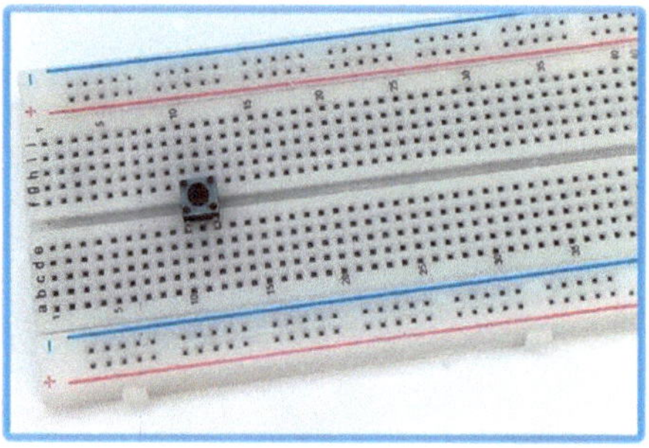

5.1

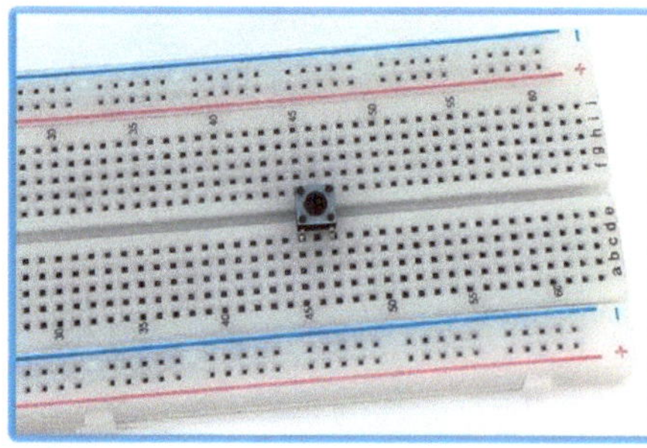

5.2

5.3

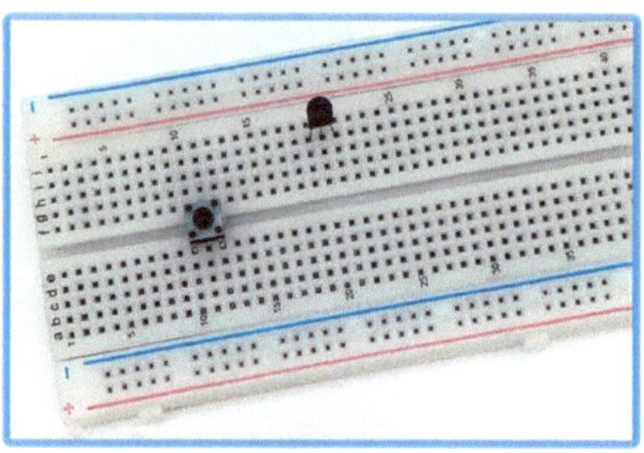

5.4

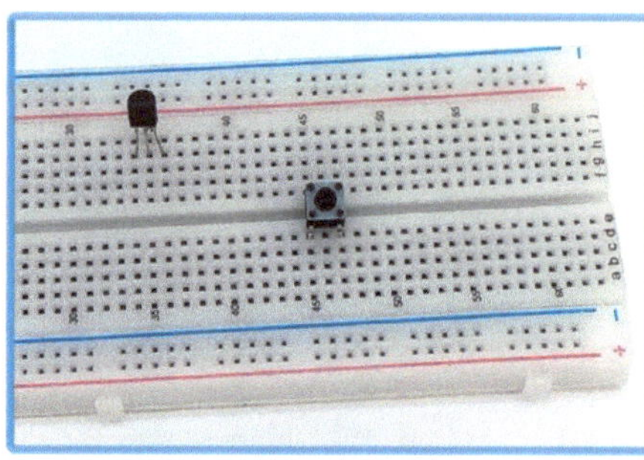

5.3: Place the collector pin of a transistor at the meeting point of i and 19 and the base pin of a transistor at the meeting point of i and 20, and the emitter pin of a transistor into the intersection of i and 21.

5.4: Place the collector pin of a transistor at the meeting point of i and 34 and the base pin of a transistor at the meeting point of i and 35, and the emitter pin of a transistor into the intersection of i and 36.

5.5: Put one end of
a 10k Ohm Resistor
pin at the meeting point of
g and 12
and the rest pin of it at
the meeting point of
g and 20.

5.5

5.6: Put one end of
a 1k Ohm Resistor
pin at the meeting point of
f and 34
and the rest pin of it at
the meeting point of
c and 34.

5.6

5.7

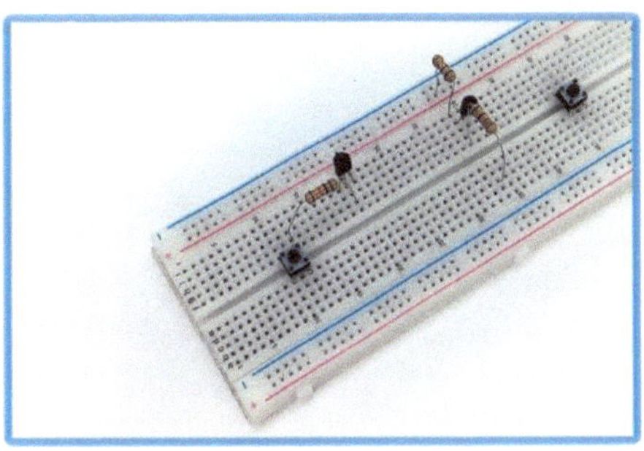

5.7: Put one end of
a 1k Ohm Resistor
pin into the junction of
j and 34
and the rest pin of it at
the positive rail
on the breadboard.

5.8

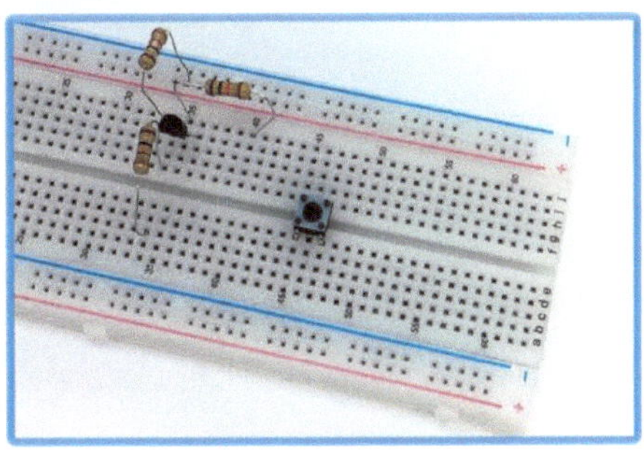

5.8: Put one end of
a 10k Ohm Resistor
pin at the meeting point of
j and 35
and the rest pin of it at
the meeting point of
j and 40.

5.9: Put the positive pin of
the blue LED
at the meeting point of
c and 50
and the negative pin of
the blue LED
into the convergence of
c and 55.

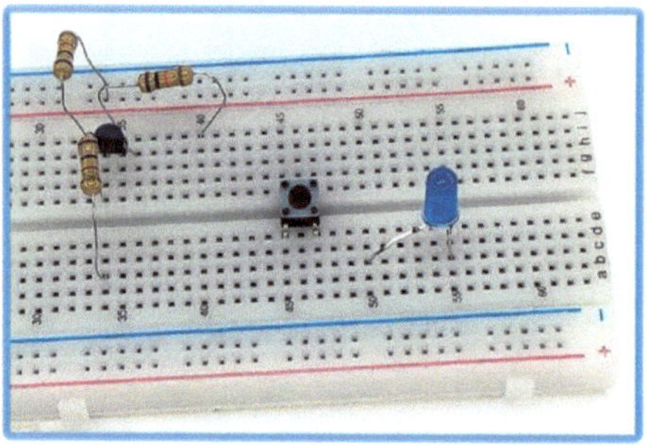

5.9

5.10: Insert one end of
a wire
into the meeting point of
j and 10
and the other end into
the positive rail
on the breadboard.

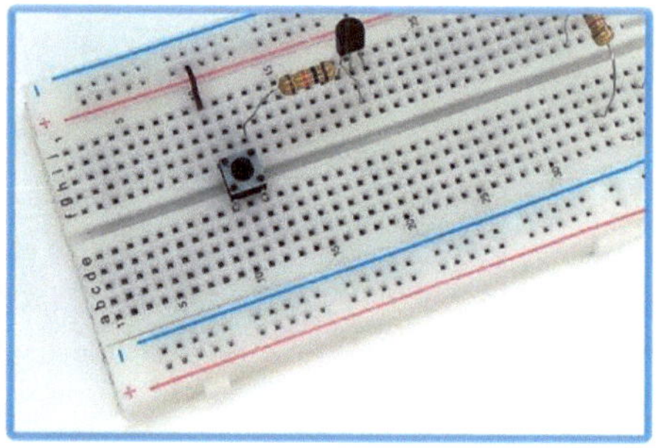

5.10

5.11

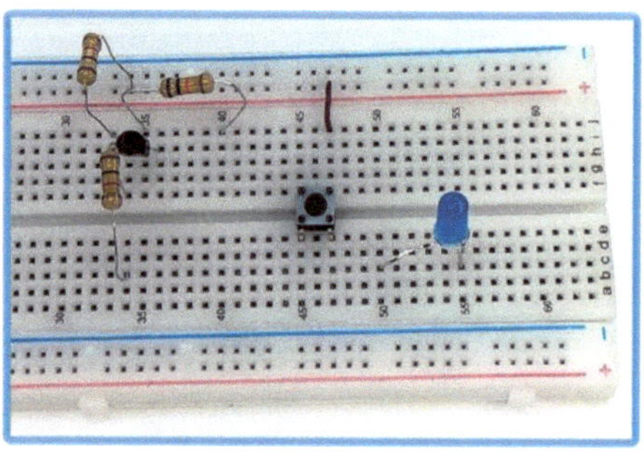

5.11: Insert one end of a
wire
into the meeting point of
j and 47
and the other end into
the positive rail
on the breadboard.

5.12

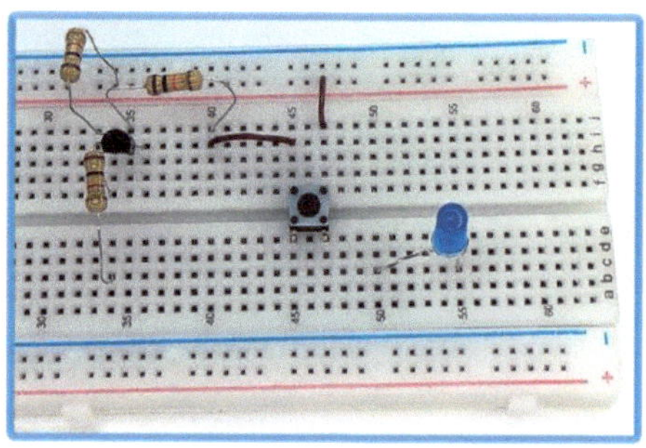

5.12: Insert one end of a
wire
into the meeting point of
i and 40
and the other end into
the meeting point of
i and 45.

5.13: Insert one end of a wire into the meeting point of h and 19 and the other end into the meeting point of h and 34.

5.13

5.14: Insert one end of a wire into the meeting point of g and 21 and the other end into the meeting point of g and 36.

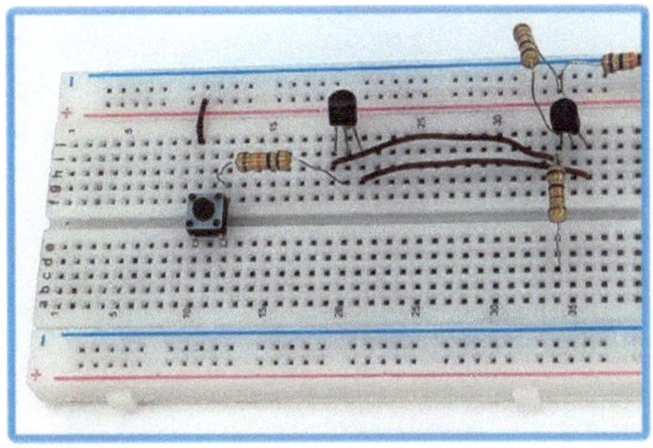

5.14

5.15

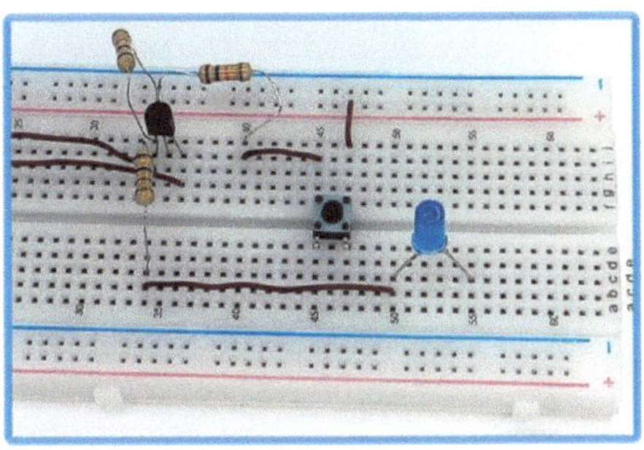

5.15: Insert one end of a wire into the meeting point of b and 34 and the other end into the meeting point of b and 50.

5.16

5.16: Insert one end of a wire into the meeting point of f and 21 and the other end into the negative rail on the breadboard.

NOR GATE

5.17: Insert one end of a wire into the meeting point of a and 55 and the other end into the negative rail on the breadboard.

5.18: Connect a 9V battery to the battery connector, and then connect the positive wire of the battery connector to the positive rail on the breadboard and the negative wire of the battery connector to the negative rail on the breadboard.

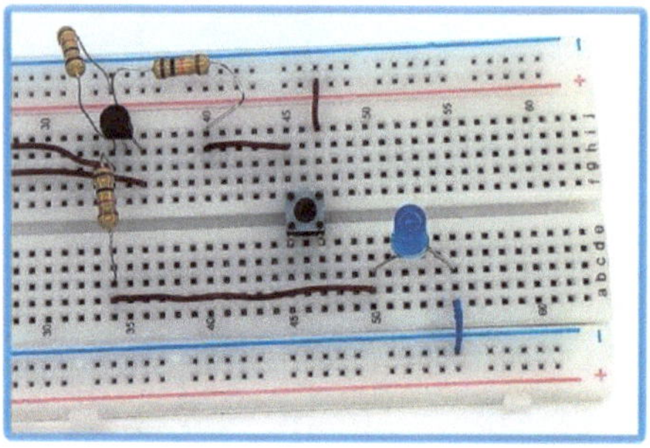

5.17

5.18

Explanation of NOR GATE

Figures (5.1) to (5.9) are the placement of components.
Figures (5.10) to (5.18) are the figures of wiring components.
The two transistors are connected in parallel.
The LED is connected to the collector of the first transistor and grounded through it.
According to the truth table of the NOR gate, if both switches are simultaneously at "OFF" state, LED will light up.
In other conditions, the LED will not light up.

PROJECT (6)

EXOR GATE

EXOR GATE

In a logic gate, multiply the inverted level of the first input level and the second input level. Then multiply the first input level and the inverse level of the second input level. A gate that combines these two sets to make a single output is called an (Exclusive OR) EXOR gate. An EXOR Gate can have two or more input levels. If the input levels of the EXOR gate are of different logic levels, the output becomes "High" or "1". If the input levels have the same logic level, the output will be "Low" or "0".

Symbol of 2-input EXOR GATE

$$Y = A \oplus B$$

Boolean Expression of EXOR GATE

INPUT		OUTPUT
A	B	A⊕B
0	0	0
0	1	1
1	0	1
1	1	0

Truth Table of EXOR GATE

Demonstration of Truth Table

1. If input A is Logic "0" and input B is Logic "0", the light will not light up as the circuit is an open circuit..

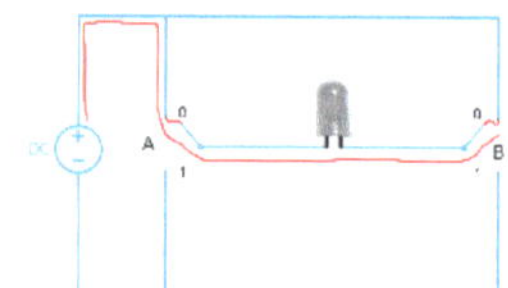

2. If input A is Logic "0", and input B is Logic "1", the light will light up as the circuit is complete.

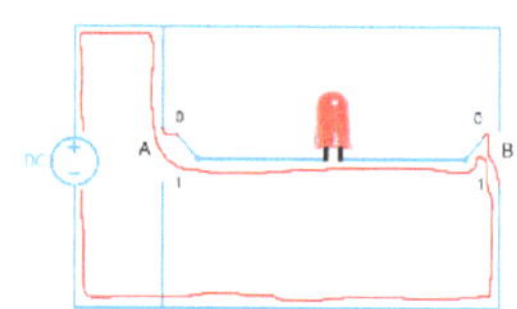

3. If input A is Logic "1", and input B is Logic "0", the light will light up as the circuit is complete.

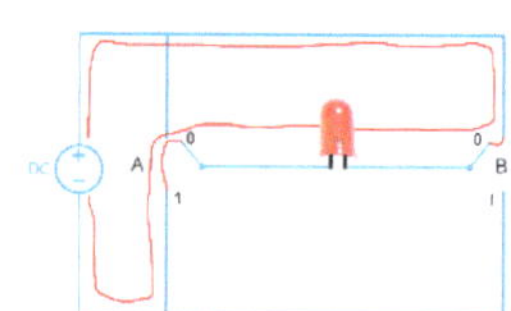

4. If input A is Logic "1" and input B is Logic "1", the light will not light up as the circuit is an open circuit.

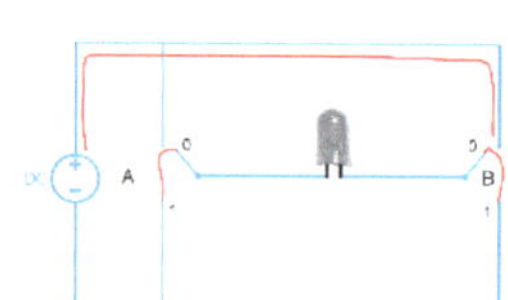

COMPONENTS

01

Push-On Switch
2 nos.

02

Transistor
4 nos.

03

Resistor 10K Ohm
4 nos.

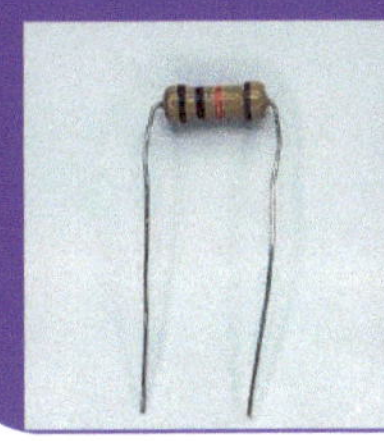

04

Resistor 1K Ohm
1 no.

05

LED
1 no.

06

Wire
13 nos.

07

Battery 9V
1 no.

08

Connector
1 no.

EXOR GATE

Steps of Implementation

6.1: Place one of the upper two pins of a Push-On Switch where f and 10 meets and the pin next to it where f and 12 intersect. Then place one of the lower two pins of that push-on switch at the meeting point of e and 10 and the pin next to it at the meeting point of e and 12.

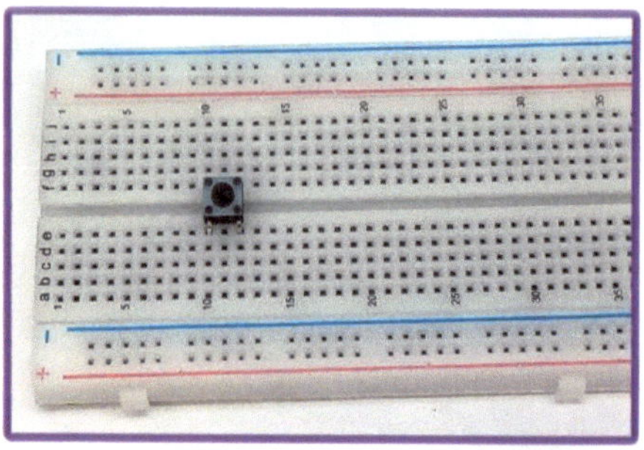

6.1

6.2: Place one of the upper two pins of a Push-On Switch where f and 15 meets and the pin next to it where f and 17 intersect. Then place one of the lower two pins of that push-on switch at the meeting point of e and 15 and the pin next to it at the meeting point of e and 17.

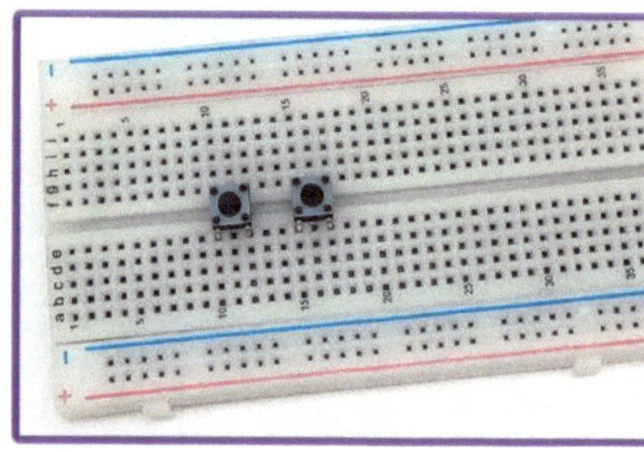

6.2

6.3

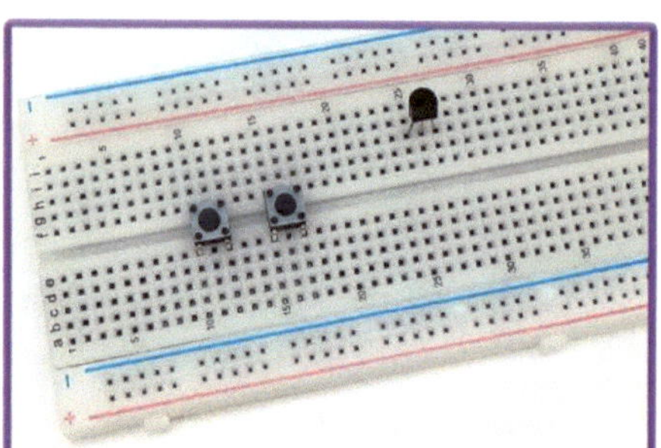

6.3: Place the collector pin of a transistor at the meeting point of h and 25 and the base pin of a transistor at the meeting point of h and 26, and the emitter pin of a transistor into the intersection of h and 27.

6.4

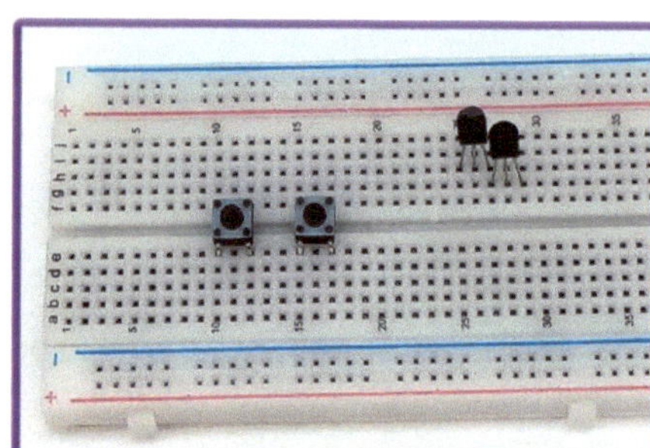

6.4: Place the collector pin of a transistor at the meeting point of g and 27 and the base pin of a transistor at the meeting point of g and 28, and the emitter pin of a transistor into the intersection of g and 29.

6.5: Place the collector pin of a transistor at the meeting point of g and 48 and the base pin of a transistor at the meeting point of g and 49, and the emitter pin of a transistor into the intersection of g and 50.

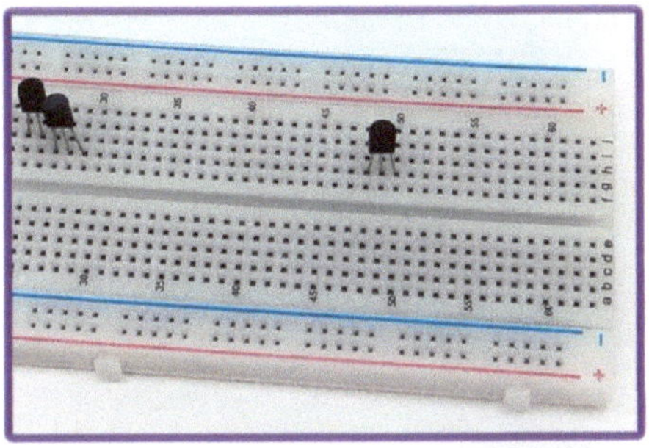

6.5

6.6: Place the collector pin of a transistor at the meeting point of g and 53 and the base pin of a transistor at the meeting point of g and 54, and the emitter pin of a transistor into the intersection of g and 55.

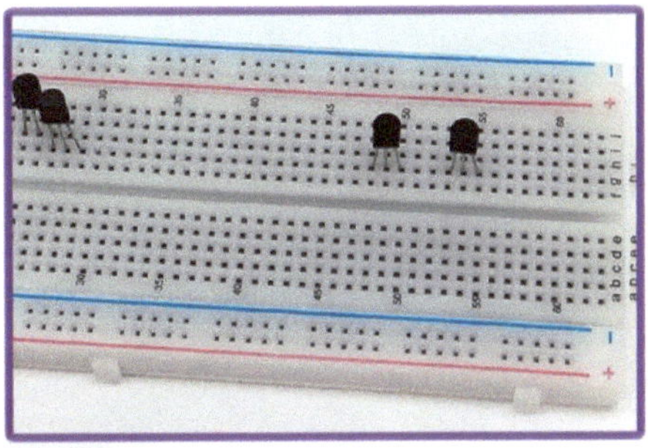

6.6

6.7

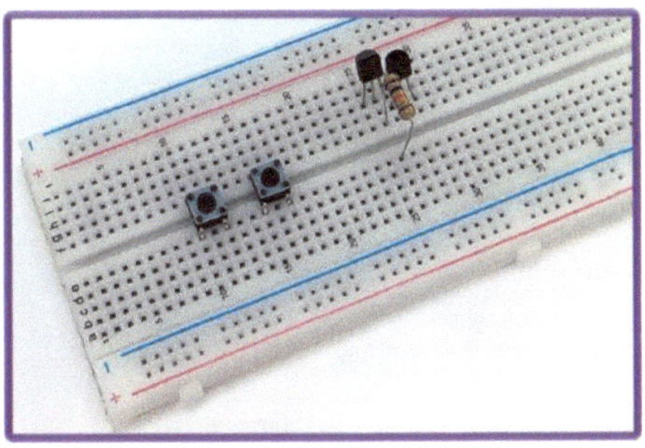

6.7: Put one end of a 10k Ohm Resistor pin at the meeting point of f and 26 and the rest pin of it at the meeting point of e and 26.

6.8

6.8: Put one end of a 10k Ohm Resistor pin at the meeting point of f and 28 and the rest pin of it at the meeting point of e and 28.

EXOR GATE

6.9: Put one end of a 10k Ohm Resistor pin at the meeting point of f and 49 and the rest pin of it at the meeting point of e and 49.

6.10: Put one end of a 10k Ohm Resistor pin at the meeting point of f and 54 and the rest pin of it at the meeting point of e and 54.

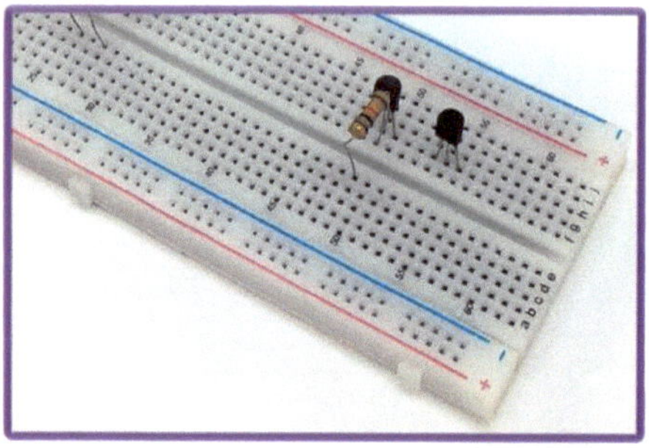

6.9

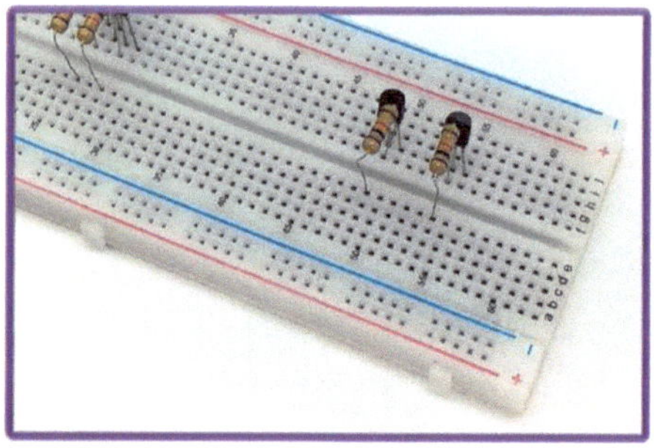

6.10

6.11

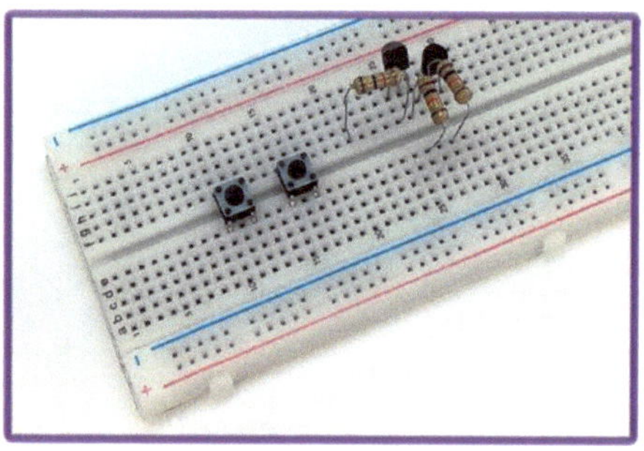

6.12

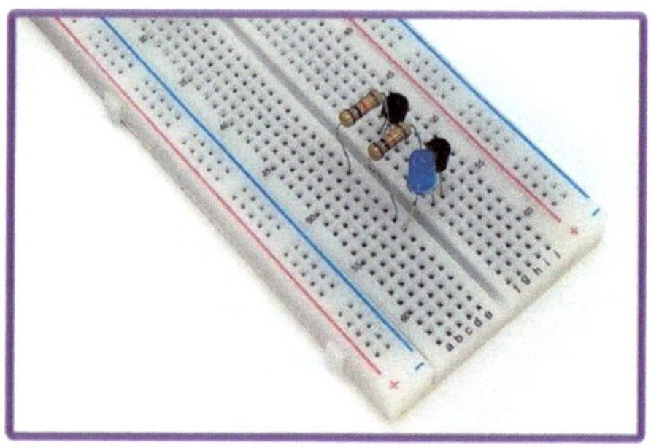

6.11: Put one end of a 1k Ohm Resistor pin at the meeting point of f and 21 and the rest pin of it at the meeting point of f and 25.

6.12: Put the positive pin of the blue LED at the meeting point of f and 55 and the negative pin of the blue LED into the convergence of e and 55.

6.13: Insert one end of a wire into the meeting point of f and 5 and the other end into the negative rail on the breadboard.

6.14: Insert one end of a wire into the meeting point of i and 5 and the other end into the meeting point of i and 29.

6.13

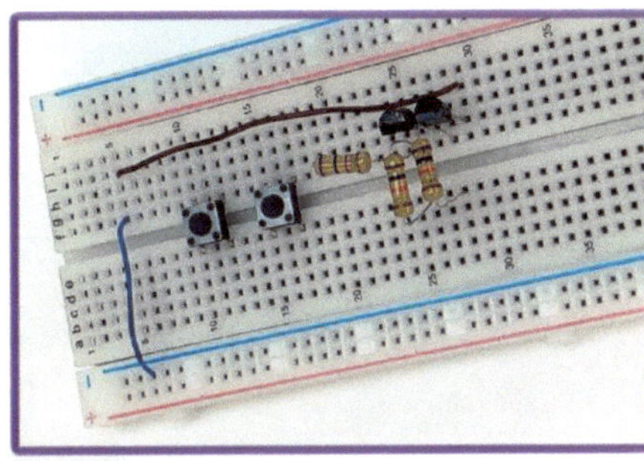

6.14

6.15

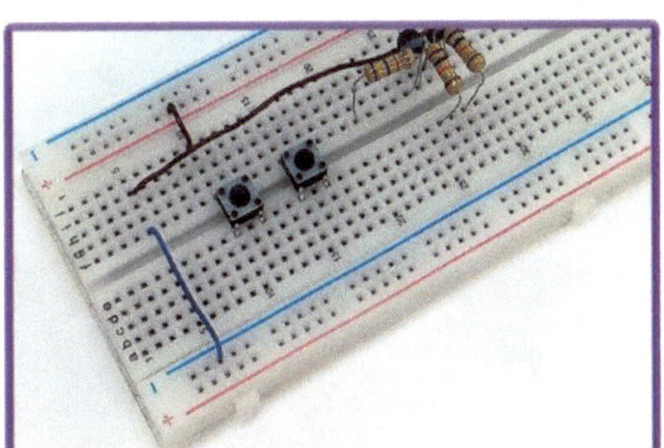

6.16

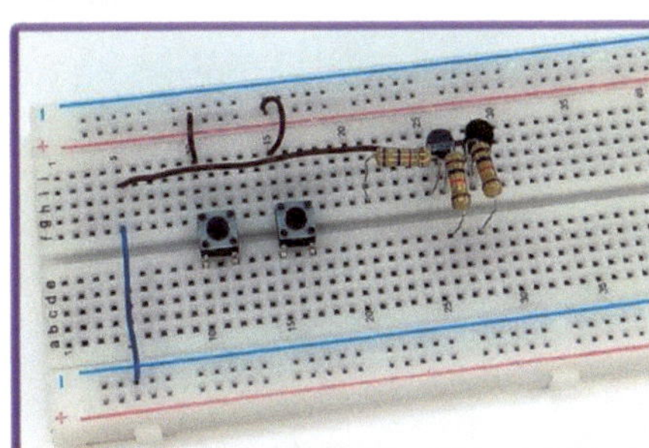

6.15: Insert one end of a wire into the meeting point of j and 10 and the other end into the positive rail on the breadboard.

6.16: Insert one end of a wire into the meeting point of j and 15 and the other end into the positive rail on the breadboard.

EXOR GATE

6.17: Insert one end of a wire into the meeting point of j and 21 and the other end into the positive rail on the breadboard.

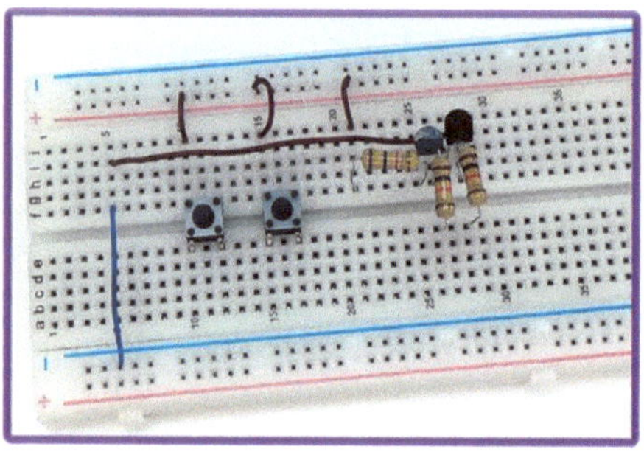

6.17

6.18: Insert one end of a wire into the meeting point of a and 55 and the other end into the negative rail on the breadboard.

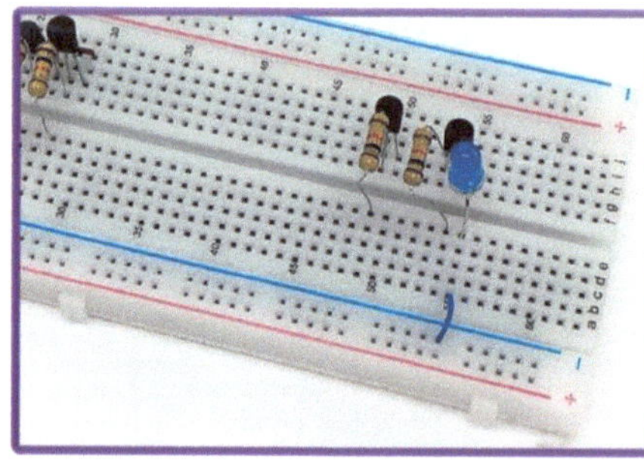

6.18

6.19

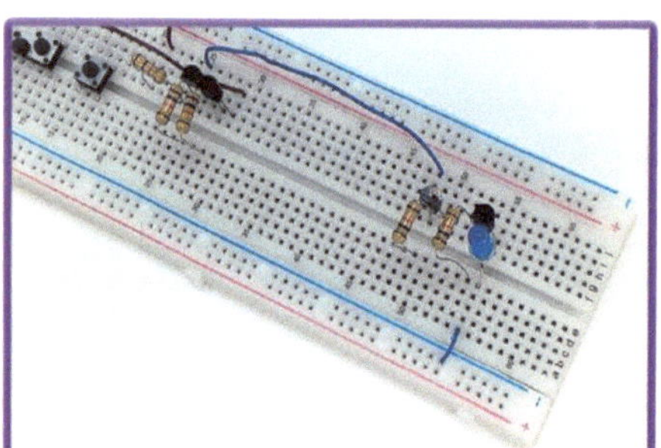

6.19: Insert one end of a wire into the meeting point of j and 25 and the other end into the meeting point of j and 48.

6.20

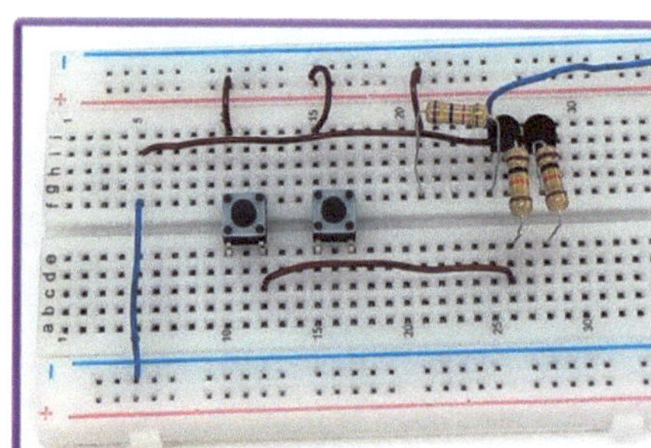

6.20: Insert one end of a wire into the meeting point of c and 12 and the other end into the meeting point of c and 26.

6.21: Insert one end of a wire into the meeting point of b and 17 and the other end into the meeting point of b and 28.

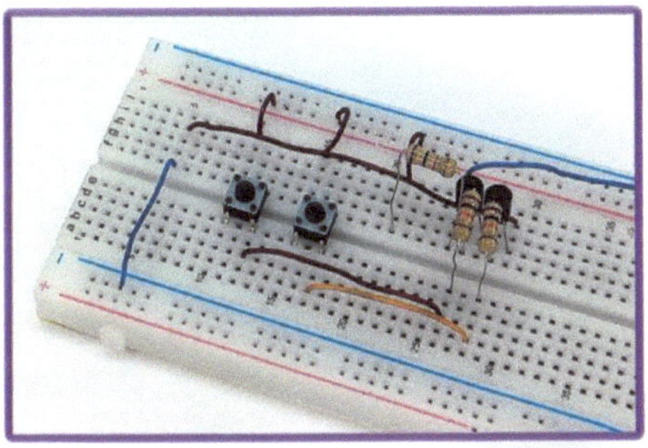

6.21

6.22: Insert one end of a wire into the meeting point of d and 26 and the other end into the meeting point of d and 49.

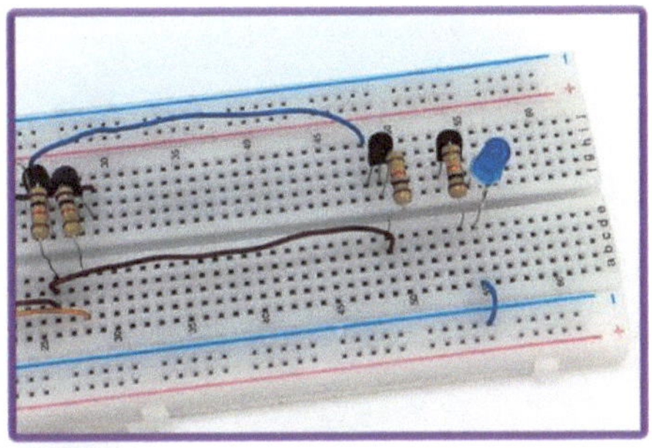

6.22

6.23

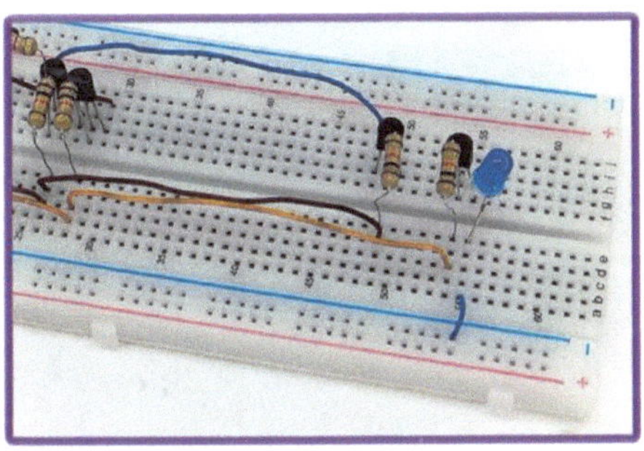

6.23: Insert one end of a wire into the meeting point of c and 28 and the other end into the meeting point of c and 54.

6.24

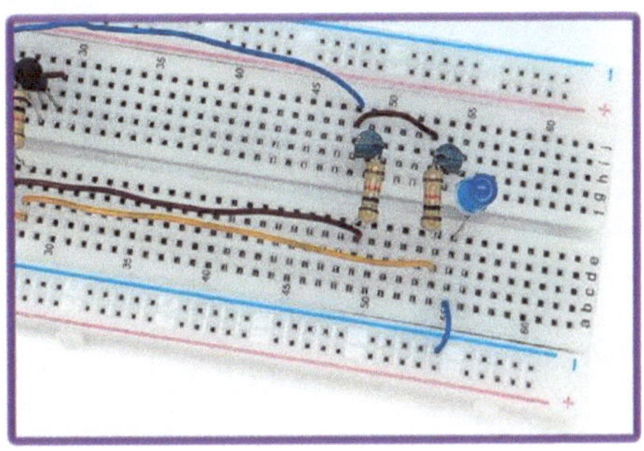

6.24: Insert one end of a wire into the meeting point of i and 48 and the other end into the meeting point of i and 53.

EXOR GATE

6.25: Insert one end of a wire into the meeting po int of j and 50 and the other end into the meeting point of j and 55.

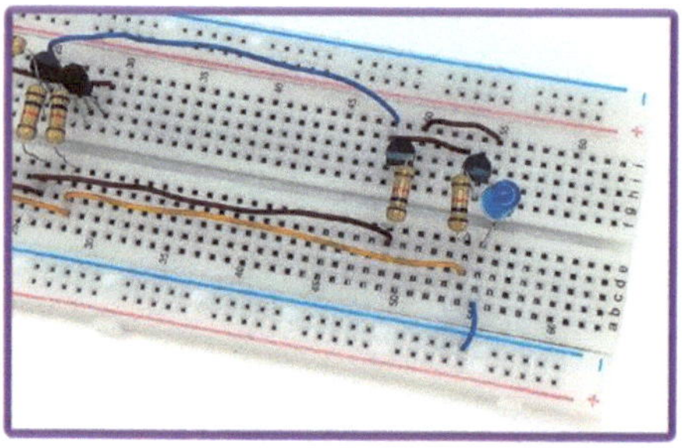

6.25

6.26: Connect a 9V battery to the battery connector, and then connect the positive wire of the battery connector to the positive rail on the breadboard and the negative wire of the battery connector to the negative rail on the breadboard.

6.26

Explanation of EXOR GATE

Figures (6.1) to (6.12) are the placement of components.
Figures (6.13) to (6.26) are the figures of wiring components.
The Two out of four transistors are connected in parallel, and the other two are connected in series.
The LED is connected to the emitter of the second of the two transistors connected in parallel and grounded through it.
According to the Truth Table of EXOR Gate, if both switches are "OFF" or "ON" state at the same time, the LED will not light up. In other conditions, the LED bulb will be on.

PROJECT (7)

EXNOR GATE

EXNOR GATE

In a logic gate, multiply the inverted level of the first input level and the second input level. Then multiply the first input level and the second input level. A gate that combines these two groups to make a single output result is called an (Exclusive NOR)EXNOR gate. EXNOR GATE is obtained by combining EXOR Gate and NOT Gate. EXNOR Gate can have two or more input levels. In EXNOR Gate, if the Input levels have different Logic levels, the output will be "Low" or "0". If the input levels have the same Logic level, the output becomes "High" or "1".

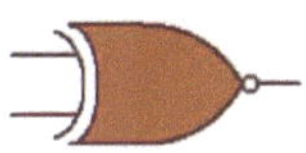

Symbol of 2-input EXNOR GATE

$$Y = \overline{A \oplus B}$$

Boolean Expression of EXNOR GATE

INPUT		OUTPUT
A	B	$\overline{A \oplus B}$
0	0	1
0	1	0
1	0	0
1	1	1

Truth Table of EXNOR GATE

Demonstration of Truth Table

1. If input A is Logic "0", and input B is Logic "0", the light will light up as the circuit is complete.

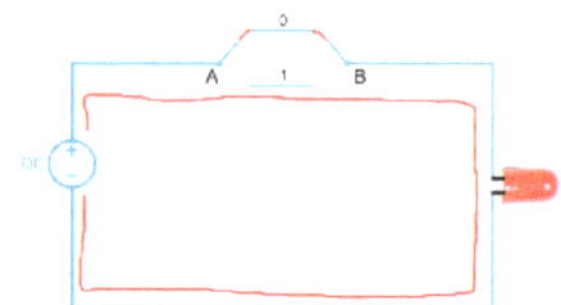

2. If input A is Logic "0" and input B is Logic "1", the light will not light up as the circuit is an open circuit.

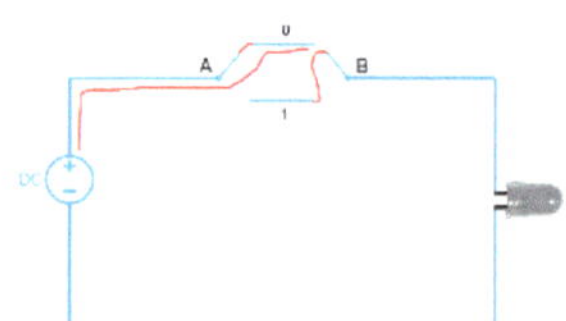

3. If input A is Logic "1" and input B is Logic "0", the light will not light up as the circuit is an open circuit.

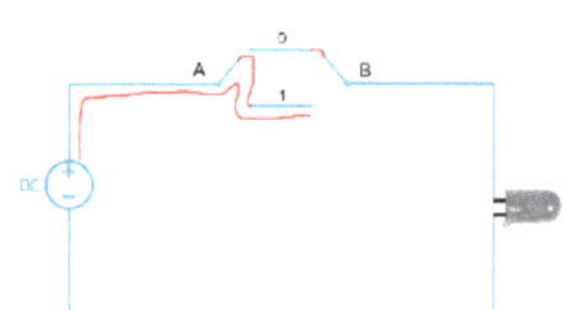

4. If input A is Logic "1", and input B is Logic "1", the light will light up as the circuit is complete.

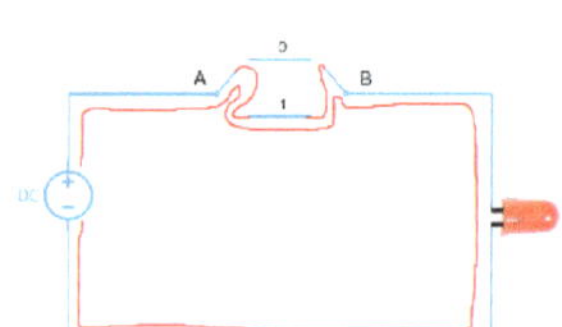

COMPONENTS

01
Push-On Switch
2 nos.

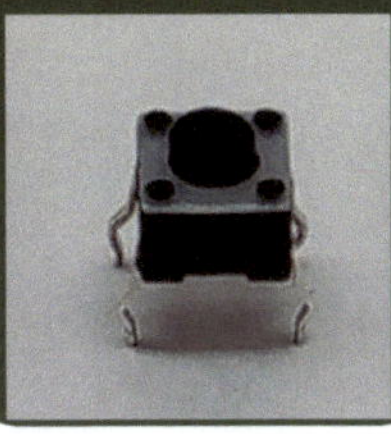

02
Transistor
5 nos.

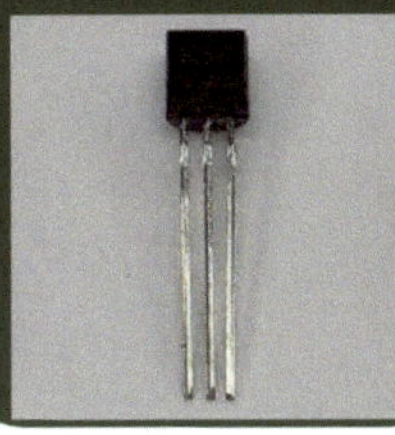

03
Resistor 10K ohm
5 nos.

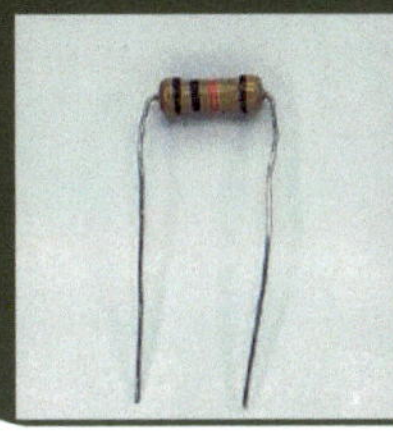

04
Resistor 1K ohm
1 no.

05
Resistor 300 ohm
1 no.

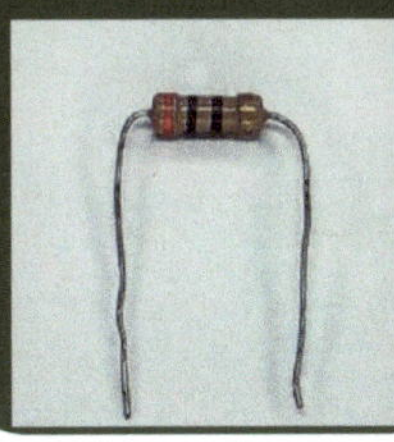

06
LED
1 no.

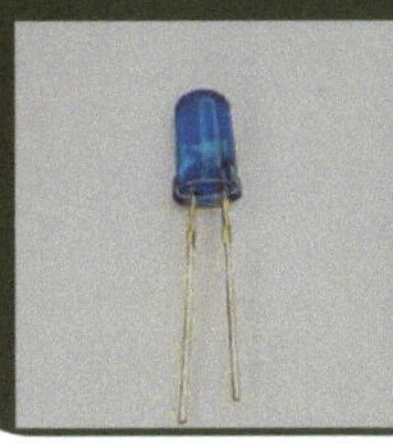

07
Wire
14 nos.

08
Battery 9V
1 no.

09
Connector
1 no.

Steps of Implementation

7.1: Place one of the upper two pins of a Push-On Switch where f and 10 meets and the pin next to it where f and 12 intersect. Then place one of the lower two pins of that push-on switch at the meeting point of e and 10 and the pin next to it at the meeting point of e and 12.

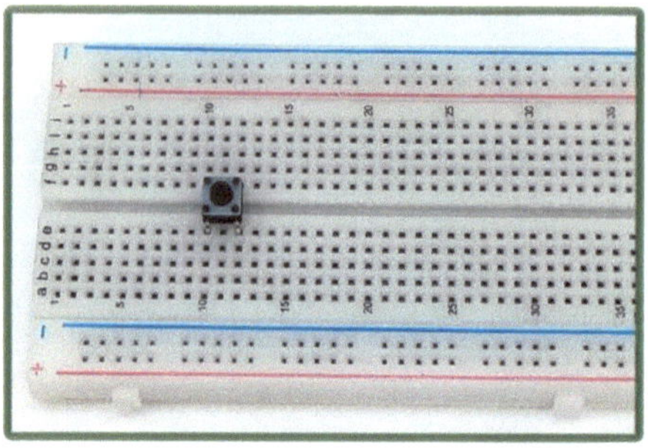

7.1

7.2: Place one of the upper two pins of a Push-On Switch where f and 15 meets and the pin next to it where f and 17 intersect. Then place one of the lower two pins of that push-on switch at the meeting point of e and 15 and the pin next to it at the meeting point of e and 17.

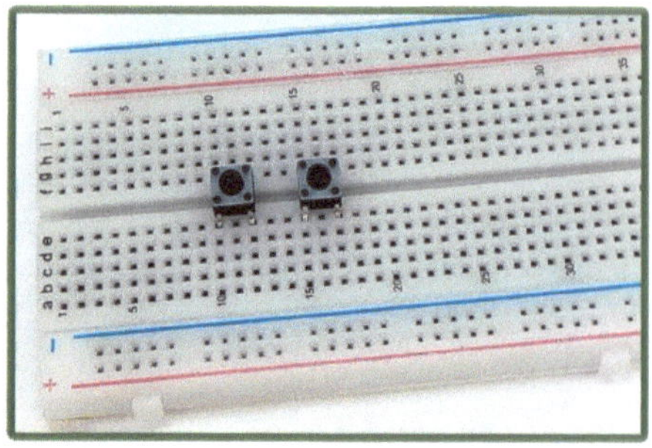

7.2

7.3

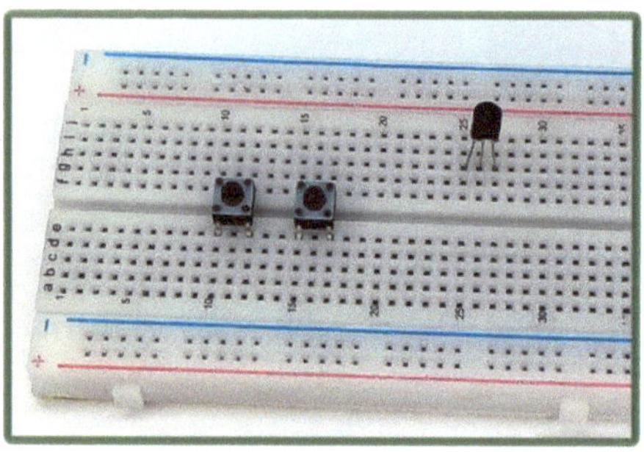

7.3: Place the collector pin of a transistor at the meeting point of h and 25 and the base pin of a transistor at the meeting point of h and 26, and the emitter pin of a transistor into the intersection of h and 27.

7.4

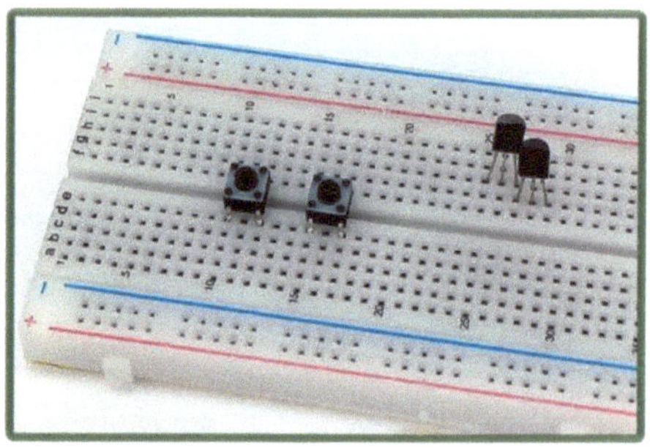

7.4: Place the collector pin of a transistor at the meeting point of g and 27 and the base pin of a transistor at the meeting point of g and 28, and the emitter pin of a transistor into the intersection of g and 29.

7.5: Place the collector pin of a transistor at the meeting point of h and 38 and the base pin of a transistor at the meeting point of h and 39, and the emitter pin of a transistor into the intersection of h and 40.

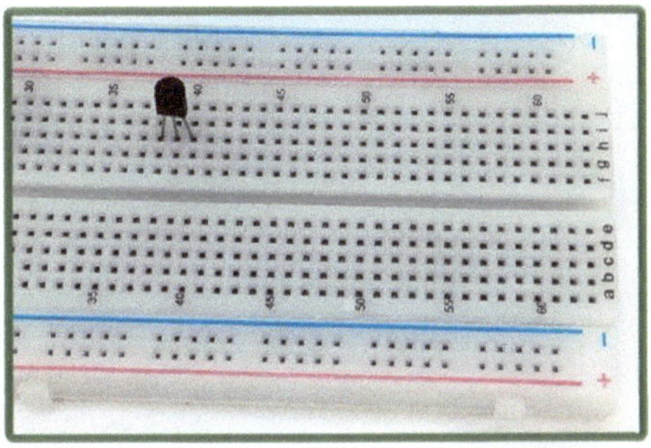

7.5

7.6: Place the collector pin of a transistor at the meeting point of h and 43 and the base pin of a transistor at the meeting point of h and 44, and the emitter pin of a transistor into the intersection of h and 45.

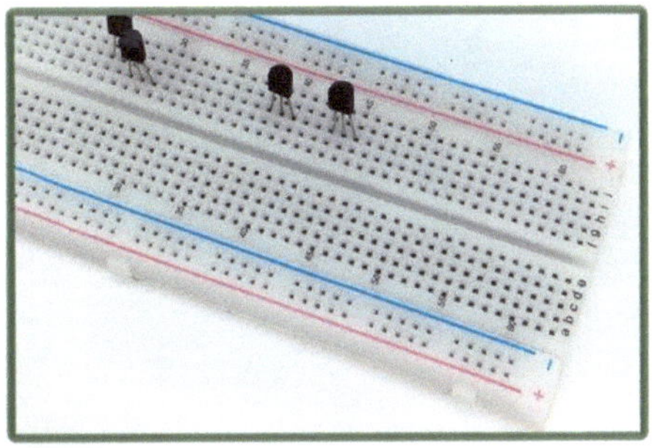

7.6

7.7

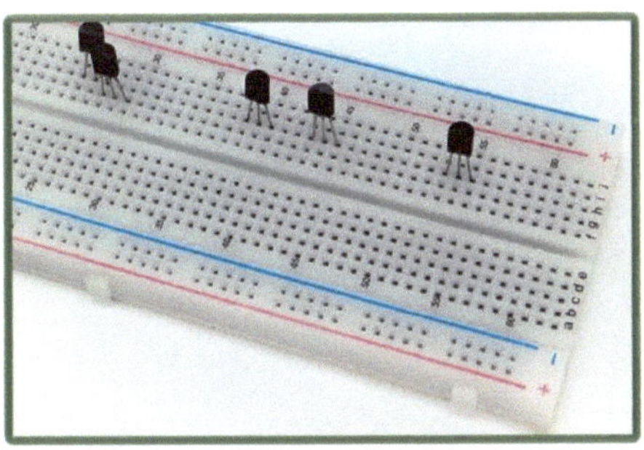

7.7: Place the collector pin of a transistor at the meeting point of h and 53 and the base pin of a transistor at the meeting point of h and 54, and the emitter pin of a transistor into the intersection of h and 55.

7.8

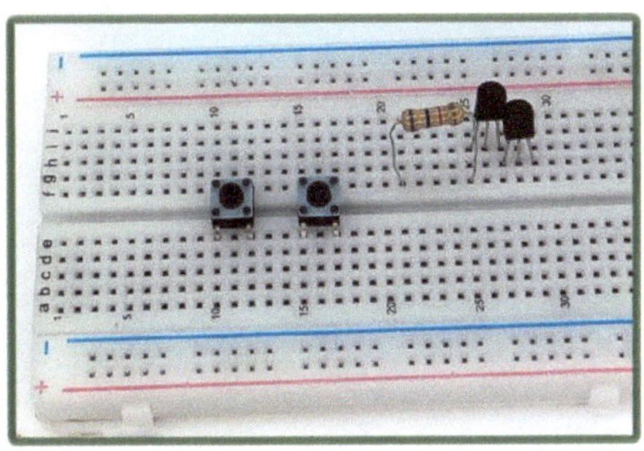

7.8: Put one end of a 1k Ohm Resistor pin at the meeting point of f and 21 and the rest pin of it at the meeting point of f and 25.

7.9: Put one end of
a 10k Ohm Resistor
pin at the meeting point of
f and 26
and the rest pin of it
at the meeting point of
e and 26.

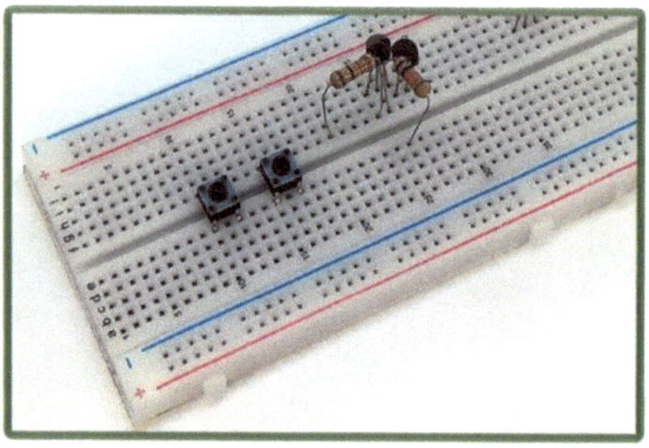

7.9

7.10: Put one end of
a 10k Ohm Resistor
pin at the meeting point of
f and 28
and the rest pin of it
at the meeting point of
e and 28.

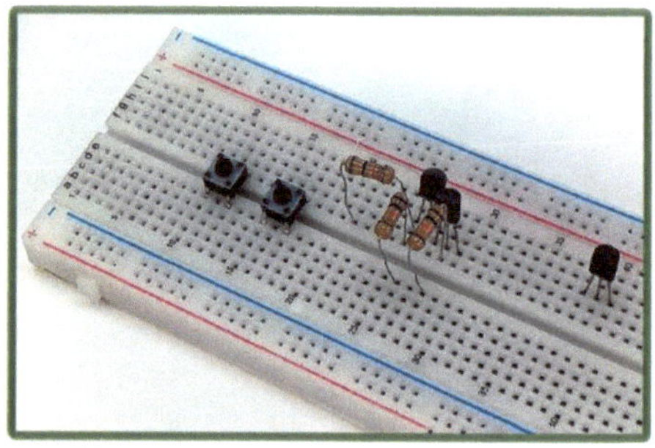

7.10

7.11

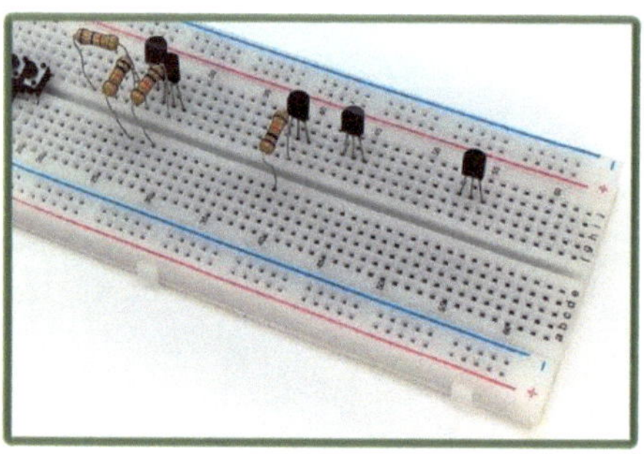

7.11: Put one end of
a 10k Ohm Resistor
pin at the meeting point of
f and 39
and the rest pin of it
at the meeting point of
e and 39.

7.12

7.12: Put one end of
a 10k Ohm Resistor
pin at the meeting point of
f and 44
and the rest pin of it
at the meeting point of
e and 44.

7.13: Put one end of
a 10k Ohm Resistor
pin at the meeting point of
g and 45
and the rest pin of it
at the meeting point of
g and 54.

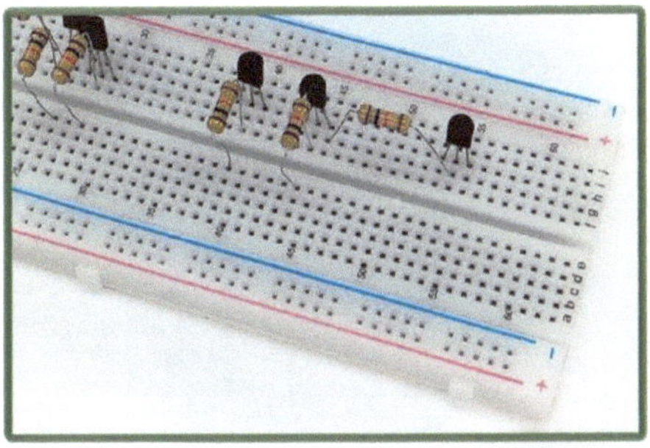

7.13

7.14: Put one end of
a 300 Ohm Resistor
pin into the junction of
j and 53
and the rest pin of it at
the positive rail
on the breadboard.

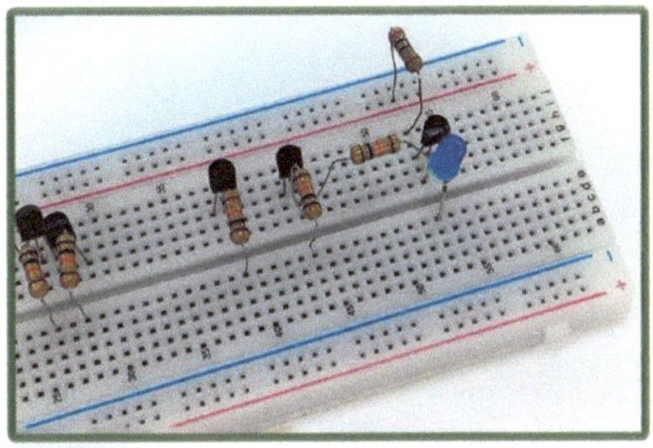

7.14

7.15

7.15: Put the positive pin of
the blue LED
at the meeting point of
f and 53
and the negative pin of
the blue LED
into the convergence of
e and 53.

7.16

7.16: Insert one end of
a wire into
the meeting point of
f and 5
and the other end into
the negative rail
on the breadboard.

7.17: Insert one end of a wire into the meeting point of i and 5 and the other end into the meeting point of i and 29.

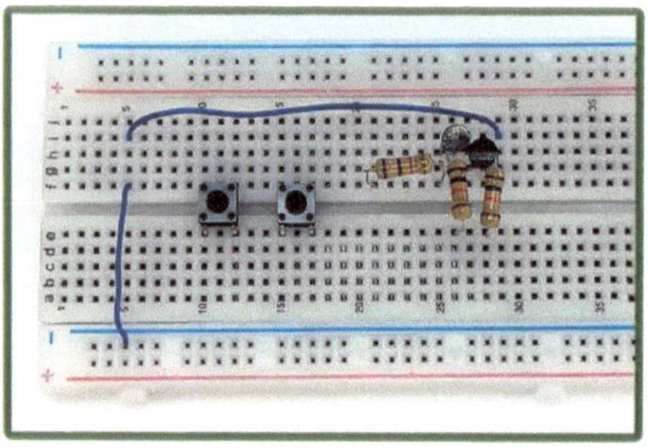

7.17

7.18: Insert one end of a wire into the meeting point of j and 10 and the other end into the positive rail on the breadboard.

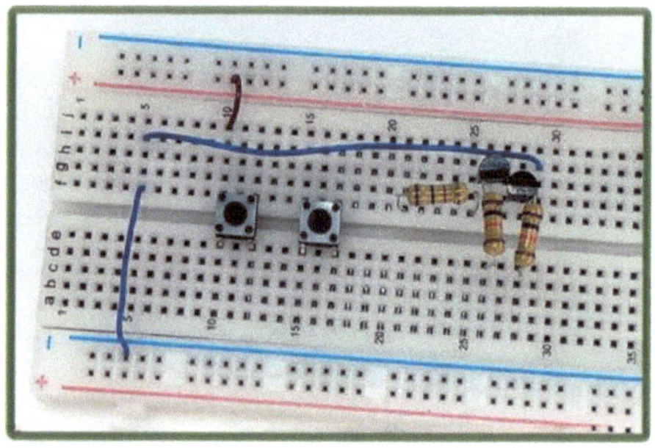

7.18

7.19

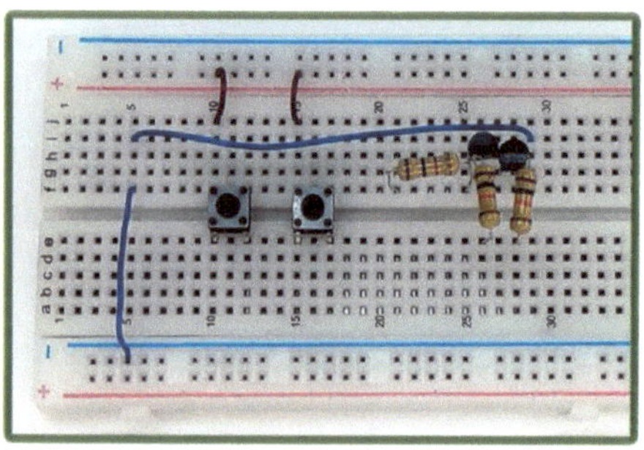

7.19: Insert one end of a wire into the meeting point of j and 15 and the other end into the positive rail on the breadboard.

7.20

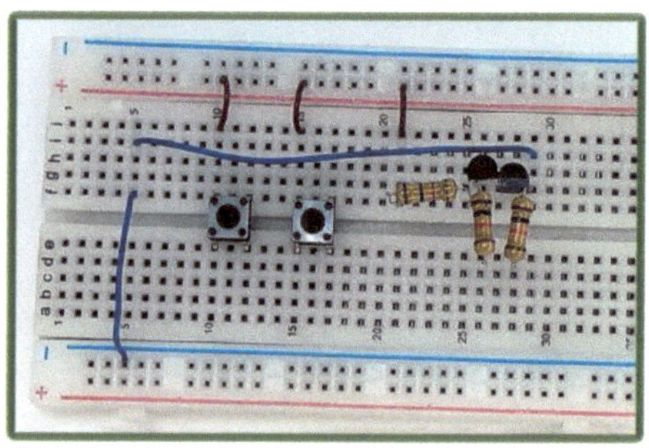

7.20: Insert one end of a wire into the meeting point of j and 21 and the other end into the positive rail on the breadboard.

7.21: Insert one end of a wire into the meeting point of c and 12 and the other end into the meeting point of c and 26.

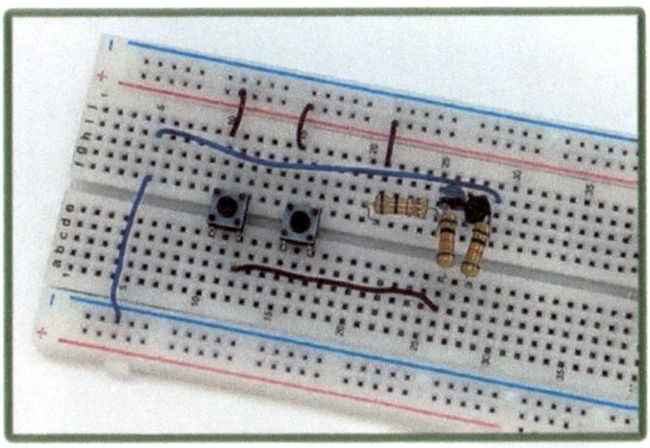

7.21

7.22: Insert one end of a wire into the meeting point of b and 17 and the other end into the meeting point of b and 28.

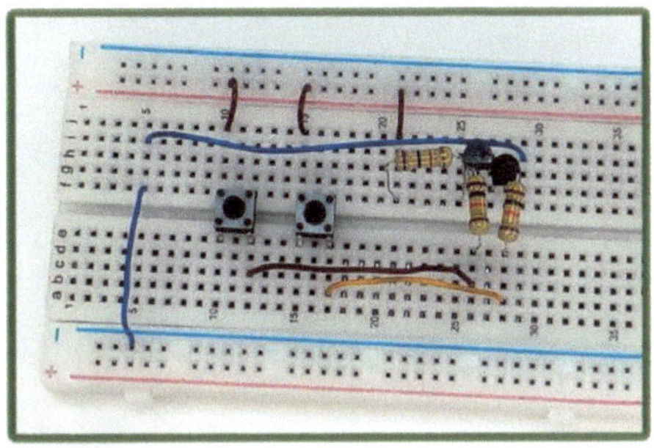

7.22

7.23

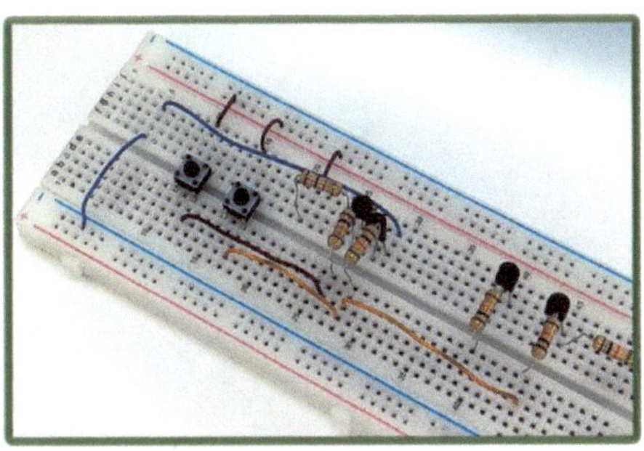

7.23: Insert one end of a wire into the meeting point of c and 28 and the other end into the meeting point of c and 44.

7.24

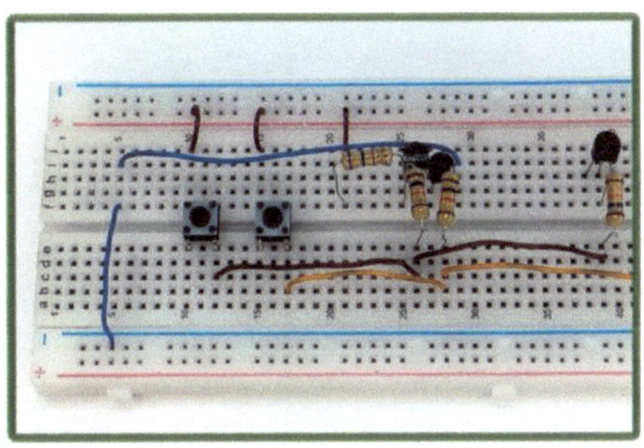

7.24: Insert one end of a wire into the meeting point of d and 26 and the other end into the meeting point of d and 39.

7.25: Insert one end of
a wire into
the meeting point of
j and 25
and the other end into the
meeting point of
j and 38.

7.26: Insert one end of
a wire into
the meeting point of
i and 38
and the other end into the
meeting point of
i and 43.

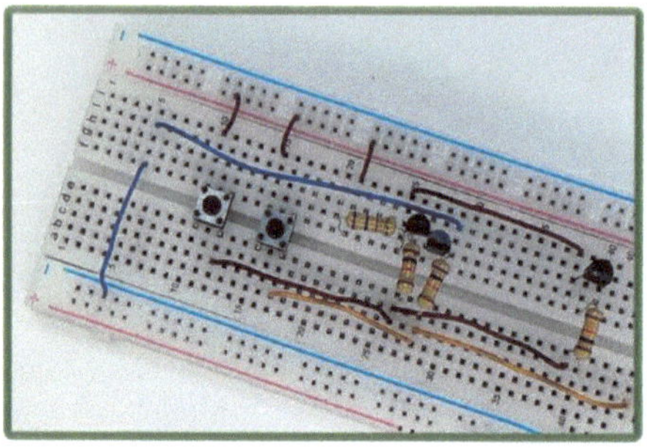

7.25

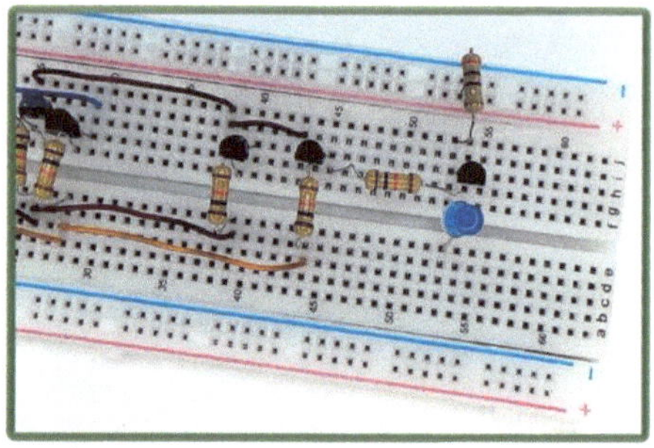

7.26

7.27

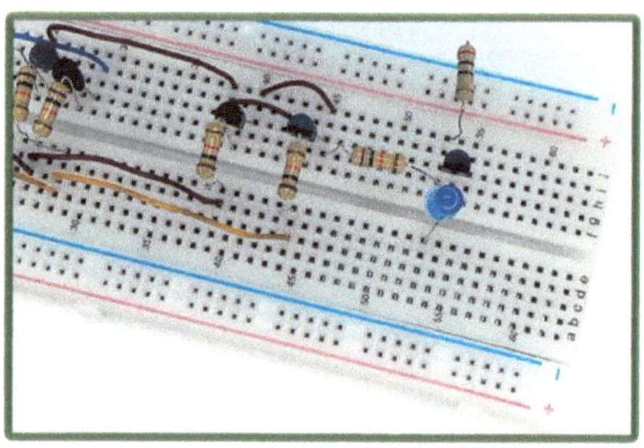

7.27: Insert one end of
a wire into
the meeting point of
j and 40
and the other end into the
meeting point of
j and 45.

7.28: Insert one end of a wire into the meeting point of a and 53 and the other end into the negative rail on the breadboard.

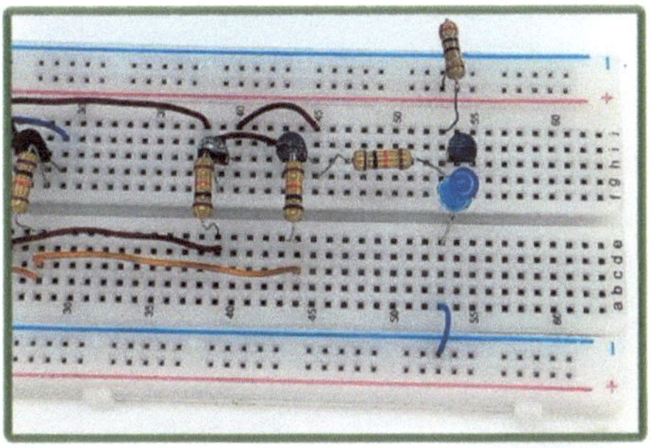

7.28

7.29: Insert one end of a wire into the meeting point of f and 55 and the other end into the negative rail on the breadboard.

7.29

7.30

7.30: Connect a 9V battery to the battery connector, and then connect the positive wire of the battery connector to the positive rail on the breadboard and the negative wire of the battery connector to the negative rail on the breadboard.

Explanation of EXNOR GATE

Figures (7.1) to (7.15) are the placement of components.
Figures (7.16) to (7.30) are the figures of wiring components.
The two out of five transistors in this circuit are connected in series, and the other two transistors are connected in parallel. The base of another transistor is connected through the emitter of the second transistor of the two parallel connected transistors.
The LED is connected to the collector of that transistor and grounded through it.
According to EXNOR Gate's truth table,
if both switches are simultaneously in at "OFF" or "ON" state, the LED will light up. In other conditions, the LED bulb will be off.

www.ingramcontent.com/pod-product-compliance
Lightning Source LLC
LaVergne TN
LVHW051136210926
848482LV00006B/332